21 CFR Part 11: Complete Guide to International Computer Validation Compliance for the Pharmaceutical Industry

21 CFR Part 11:
Complete Guide to
International Computer
Validation Compliance for
the Pharmaceutical Industry

Orlando López

CRC Press
Taylor & Francis Group
Boca Raton London New York

CRC Press is an imprint of the
Taylor & Francis Group, an **informa** business

CRC Press
Taylor & Francis Group
6000 Broken Sound Parkway NW, Suite 300
Boca Raton, FL 33487-2742

First issued in paperback 2019

© 2004 by Taylor & Francis Group, LLC
CRC Press is an imprint of Taylor & Francis Group, an Informa business

No claim to original U.S. Government works

ISBN-13: 978-0-8493-2243-3 (hbk)
ISBN-13: 978-0-367-39459-2 (pbk)

Library of Congress Card Number 2003063460

Library of Congress Cataloging-in-Publication Data

López, Orlando.
  21 CFR Part 11 : complete guide to international computer validation compliance for the pharmaceutical industry / Orlando López.
      p. cm.
  Includes bibliographical references and index.
  ISBN 0-8493-2243-X (alk. paper)
   1. Pharmaceutical industry. 2. Computer software—Validation, 3. Computer programs—Verification. I. Title: Twenty-one CFR Part Eleven. II. Title.

RS192.L67 2004
338.4'76151—dc22

2003063460

**Visit the Taylor & Francis Web site at
http://www.taylorandfrancis.com**

**and the CRC Press Web site at
http://www.crcpress.com**

Contents

Foreword by Sion Wyn.. ix
Preface by Orlando López xi
Publisher's Note... xiii
Dedication... xv

Chapter 1   Introduction 1

Chapter 2   Validation Overview.. 5
What Is a Computer System?.. 5
What Is a Computer Systems Validation?.................................... 5
Why Do We Validate Computer Systems? 6
Key Project Elements 8
Which Systems Should Be Validated? 8
Introduction to the Computer Systems Validation Process 9
Computer Systems Validation for Low Criticality and/or Low
Complexity Projects.. 11

Chapter 3   USA Regulatory Requirements for Computer Systems 13
Medical Devices Software 17
The Food Industry 18

Chapter 4   New Computer Systems Validation Model 21

Chapter 5   Computer Validation Management Cycle....................... 25
Validation Policies.. 26
Validation Guidelines.. 26
Validation Plans... 27
Procedural Controls.. 27
Compliance Assessments... 27
Validation of Computer Systems 27
Supplier Qualification 27
Ongoing Support Systems 27

Chapter 6   Computer Validation Program Organization 29
Organizational Model 29
Computer Systems Validation Executive Committee 30
CSV Cross-Functional Team 30
CSV Groups and Teams... 31
The Management Group 32
Validation Program Coordinators.. 32

Chapter 7   **The Computer Systems Validation Process** 35
                 System Development Files 40

Chapter 8   **Validation Project Plans and Schedules** 43
                 Regulatory Guidance 43
                 Validation Project Plans 43
                 Mandatory Signatures 45
                 Project Schedule 45

Chapter 9   **Inspections and Testing** 49
                 Regulatory Guidance 49
                 Introduction 49
                 Document Inspections and Technical Reviews 50
                 White Box Testing 51
                 Black Box Testing 52
                 Other Testing Types 54

Chapter 10   **Qualifications** 57
                 Introduction 57
                 Hardware Installation Qualification 58
                 Software Installation Qualification 61
                 System Operational Qualification 64
                 System Performance Qualification 67
                 Operating System and Utility Software Installation Verification 69
                 Standard Instruments, Microcontrollers, Smart Instrumentation
                 Verification 70
                 Standard Software Packages Qualification 73
                 A Related Product for ISO/IEC 12119, The IEEE
                 Standard Adoption of ISO/IEC 12119 73
                 Configurable Software Qualification 76
                 Custom-Built Systems Qualification 78

Chapter 11   **SLC Documentation** 81
                 Regulatory Guidance 81
                 SLC Documentation 81

Chapter 12   **Relevant Procedural Controls** 85

Chapter 13   **Change Management** 87
                 Introduction 87
                 Change Management Process 88

Chapter 14   **Training** 91
                 Regulatory Guidance 91
                 Training in the Regulated Industry 91

Chapter 15   **Security** 93
                 Regulatory Guidance 93
                 Introduction 93

Physical Security 96
Network Security 97
Applications Security 98
Other Key Security Elements 99

Chapter 16 Source Code 105
Regulatory Guidance 105
Introduction 105

Chapter 17 Hardware/Software Suppliers Qualification 107

Chapter 18 Maintaining the State of Validation 111
Security 111

Chapter 19 Part 11 Remediation Project 117
Introduction 117
Evaluation of Systems 118
Corrective Action Planning 119
Remediation 119
Remediation Project Report 120

Chapter 20 Operational Checks 121
Instructions to Operators 121
Operation Sequencing 121
Part 11-Related Operational Checks 122
Validation of Operational Checks 124

Chapter 21 Compliance Policy Guide (CPG) 7153.17 125
Introduction 125

Chapter 22 Electronic Records 129
Regulatory Guidance 129
What Constitutes an Electronic Record? 129
What Constitutes a Part 11 Required Record? 130
How Should Part 11 Records Be Managed? 130
Minimum Record Retention Requirements 131
When Are Audit Trails Applicable for Electronic Records? 131
Instructions 132
Events 132
Reviews 133
Preservation Strategies 133
Electronic Records Authenticity 134
Storage 135

Chapter 23 Electronic Signatures 137
Regulatory Guideline 137
General Concepts 137
Password-Based Signatures 138
Digital Signatures 138

Chapter 24   Technologies Supporting Part 11 141
            Paper-Based versus Electronic-Based Solutions 141
            Hash Algorithms 142
            Data Encryption. 142
            Digital Signatures 145
            Windows® OS 145

Chapter 25   All Together 147
            Acquisition Process 147
            Supply Process. 148
            Development Process. 148
            Operation Process 150
            Maintenance Process 150

Chapter 26   The Future 153

Appendices
A   Glossary of Terms. 157
B   Abbreviations and Acronyms 165
C   Applicability of a Computer Validation Model 167
D   Criticality and Complexity Assessment. 173
E   Sample Development Activities Grouped by Project Periods. 183
F   Administrative Procedures Mapped to Part 11 209
G   Sample Audit Checklist for a Closed System 215
H   Computer Systems Regulatory Requirements 219
I   Technical Design Key Practices 239

Index 241

Foreword

It has been my pleasure to know Orlando López as a friend for some years, and it has also been my privilege to work with him in his specialty field, computer systems validation and compliance. Orlando is well known in the industry as an enthusiastic advocate and supporter of industry initiatives such as GAMP, and a valued contributor to various task teams and committees.

He is a practitioner with valuable experience in tackling real challenges and dealing with real-life problems. In this book, he illuminates the role of quality assurance, and shows the importance of integrating the validation activities into the system life cycle within a structured top-down approach.

This book reflects Orlando's breadth of knowledge of and experience in the regulated pharmaceutical and associated healthcare industries, software engineering, and quality assurance. It covers the regulatory requirements, and the organisation, planning, verification, and documentation activities required to meet those requirements. It also describes the administrative and procedural controls required for compliance, as well as introducing appropriate supporting technologies such as encryption and digital signatures.

Orlando is enthusiastic about his subject, and he is a mine of useful up-to-the-minute information. He is always ready and willing to share this information and experience with others for the benefit of the industry and the patient.

This book shows his commitment to the practical application of quality assurance and engineering techniques in the development of systems that meet user and regulatory requirements.

Sion Wyn
Conformity Limited
United Kingdom

Preface

This book speaks to all those involved in planning and undertaking computer systems validation in the United States of America's Food and Drug Administration (FDA) regulated industries, mainly pharmaceuticals. The topics and discussion levels reflect what I believe such people will find interesting or useful. My vision for this book is that it reveals my perspective on how to perform computer systems validation, and gives my readers a clear picture on how they must work effectively in a quality role in today's high-pressure, regulated environment.

Focusing on project management, I have chosen to include materials not normally stressed in a computer systems book applicable to the FDA regulated industries. Estimating, planning, and scheduling computer systems validation are hard to pin down, because there is no true end to a project. More verification and testing are always required, and there are serious risks associated with skipping them. Project managers must be fully aware of these 'trade-offs,' and need to share this knowledge with their teams. Efficiency is a major concern in this book.

Computer systems validation personnel must also deal with design errors. A program that perfectly meets a lousy specification is a lousy program. Specifically for medical devices, books on software reliability tend to set aside the user interface issue, and treat it as the sole province of the human factor analyst. The reliability of a system is determined by how all its various parts, including the people who use it, work together.

Your role as a project manager is to find and highlight problems in a product, with the aim of improving its quality. You are one of the few who will have the required information, or be directly able to examine the product in close detail, before it is transferred to the regulated environment sector.

In the FDA-regulated environment, it is not acceptable for programmers or their managers to deviate from standardized methodologies. Control programs for manufacturing facilities must be well documented and thoroughly specified. Computer systems must only change after careful control, since the electronic records associated within these systems are now highly relevant to new FDA regulations. If manufacturing systems fail, the quality of the product is in question: this can be catastrophic to the users, patients whose lives depend largely on quality assurance staffs, and on huge computer systems budgets being met.

It is fascinating how, in many highly standardized environments, a computer systems validation (CSV) group works with the product just before it goes to production, or when it has just been put into internal use.

This group is not involved in the drug development effort, their budgets are often tiny, and their deadlines almost impossible to achieve. The quality assurance organization may 'look down its collective nose' at them and the CSV team is not involved with a product for the time required to perform 'real' product life-cycle verification and testing.

With the introduction of 21 CFR Part 11 by the FDA, CSV practitioners are encountering a new wave of computer systems validation. The computer validation professional needs to be aware of quality engineering, project management, and computer technologies. This book emphasizes the promotion of a conceptual understanding of computer systems validation, and the need for evidence of such understanding in the form of practical examples.

Commencing with a discussion of the regulatory requirements associated with the practical foundation work for any project, this book introduces a model to determine the Part 11 requirements for consideration during implementation; this is based on essential characteristics of computer systems. This book then establishes a top-down approach to the integration of computer systems validation strategies in support of a computer systems validation project; focussing on the practical issues in computer systems implementation and operational life project management; including relevant activities to comply with Part 11, validation planning, and scheduling. Inspections and testing, computer systems qualification, and sample qualification using the Good Automated Manufacturing Practice (GAMP) categories of software, are all covered here.

The second section reviews documentation in relation to CSV and the procedural controls required for regulated operations. Change management and control, training, and security (introducing key technology-driven services such as user/data authentication and access control), are followed by guidance on source code issues, and suppliers' qualifications.

The book explains how technologies such as hashing, encryption, and digital signatures can support Part 11. These technologies have not currently been implemented to any extent in the FDA-regulated environments.

Finally, my vision of the integration of project management and computer systems validation is based on hands-on experience and the knowledge accumulated from more than 20 years experience in software engineering and software quality engineering. The author wishes to acknowledge and express appreciation to all of the people who, during the last 30 years, have contributed to this body of knowledge, and who laid down the crucial groundwork for this computer systems validation book.

Orlando López, Executive Consultant
J&J Networking and Computing Services
Raritan, New Jersey

Publisher's Note

The FDA withdrew its last guidance document in the series on electronic copies of electronic records in February 2003. But it did NOT withdraw the 21 CFR Part 11 Rule, which is still in force, and must be complied with: the FDA is still inspecting companies for compliance and none of the multi-million dollar fines imposed by the FDA for non-compliance have been revoked. Nor have any of the preceding warning letters been recalled.

This Professional Edition of a pragmatic guide is designed to enlighten the industry at large, which suffers somewhat from the confusion created by earlier guidance documents now withdrawn. Written by an international expert in computer software and computer validation, it will become a classic of its kind when properly used to help companies implement their compliance requirements to meet the FDA and the MCA.

On September 3, 2003 the FDA released the final guidance on the 'Scope and Application' of 21 CFR Part 11. Comments from industry played a large role in the changes that the FDA made between the draft version and the new final version. Significant clarification was also established on the definition of legacy systems and on 'enforcement discretion'.

Despite these clarifications, there is still a great deal of uncertainty remaining regarding what is expected by the Agency in the long term and how to proceed logistically with a risk-based remediation for Part 11. To access these guidelines, please visit the following website: http:/www.fda.gov/cder/guidance/5667fnl.htm

For Lizette,
Mikhail, István, and Christian

Chapter 1

Introduction

FDA regulations require that domestic or foreign manufacturers of regulated products intended for commercial distribution in the United States, establish and follow controls as part of a quality assurance (QA) program. These controls help to ensure that regulated products are *safe and effective*. The QA controls and associated records required by the FDA are contained in the applicable predicate regulations.

The records-keeping requirements to demonstrate the safety and efficacy of products are also applicable for persons exporting human drugs, biological products, devices, animal drugs, food, and cosmetics that may not be marketed or sold in the United States.[1]

Computer systems used to collect and manage records must demonstrate to the FDA that regulated products are safe and efficacious. Increased attention being paid to computers systems by the regulatory authorities requires that a periodic evaluation of compliance must be carried out by computer systems developers,[2] companies, industry, and/or the regulator.

As with any production equipment performing regulated operations, computer systems must be designed and deployed in compliance with specific performance and quality standards. In the FDA regulation context, validation is the process of proving, in accordance with the principles of the applicable predicate regulations, that any procedure, process, equipment, materials, activity, or system actually leads to the expected result. The documentation generated during the validation can be subject to examination by FDA field investigators. The results of a high-quality validation program can ensure a high degree of assurance of the trustworthiness of the electronic records and computer system functionality.

Computer systems validation, as established in 21 CFR Part 211.68, Automatic, Mechanical, and Electronic Equipment, is one of the most important requirements in FDA-regulated operations and an element of the system life cycle (SLC). In addition to the testing of the computer technology, other verifications and inspection activities include code walkthroughs, dynamic analysis and trace analysis. These activities may require 40% of overall project efforts.

In common with all FDA-regulated products, quality is built into a computer system during its conceptualization, development, and operational life. The validation of computer systems is an ongoing process that is integrated into the entire SLC.

Given the type of software commonly found in FDA-regulated operations, the criticality and complexity of computer systems, the evolving regulatory climate, and the current industry best practices, no single fixed standard can be applied to computer systems validation. However, 21 CFR Part 11; Electronic Records, Electronic Signatures Rule (hereafter referred to as Part 11), effective since August 1997, provides the explicit and current regulatory trends applicable to computer systems performing regulated operations. Only electronic records and electronic signatures that meet Part 11 can be used to satisfy the requirements in the applicable predicate rule.

---

[1] Federal Register: December 19, 2001 (Vol. 66, No. 244).

[2] Note. In this book a 'developer' can either be an external company or an in-house software development group.

1

A computer systems validation program for regulated operations can be established based on the regulatory requirements, SQA practices, SQE practices, and type of software.

This book, relevant to FDA-regulated operations, provides practical information to enable compliance with computer systems validation requirements, while highlighting and efficiently integrating the Part 11 requirements into the computer validation program. The ideas presented in this book are based on many years of experience in the United States Department of Defense and FDA-regulated industries in various computer systems development, maintenance, and quality functions.

The practical approach we take is intended to increase efficiency and ensure that correct software development and maintenance is achieved. The topics addressed in this book are equally relevant to other type of software development industries. It is not the intention of this book to develop a paradigm or model for the regulated industry.

Chapter 2 introduces the reader to the computer systems validation concepts in the context of regulated operations.

Chapter 3 discusses the regulatory requirements and furnishes the reader with the foundations of the CSV approach in this book.

Chapter 4 introduces a model to help determine the Part 11 requirements to be considered during implementation. It focuses on the basic characteristics of computer systems. Practical examples on how to implement this model are provided in Appendix C.

Chapter 5 establishes the top-down approach for integrating computer systems validation strategies in support to the computer systems validation program.

Chapter 6 suggests an organization structure supporting the computer validation program.

Chapter 7 is the focus of this book, being a practical approach to computer systems implementation and operational life project management, and including the activities relevant for compliance with Part 11. Appendix E expands the concept presented in this chapter.

Chapter 8 covers validation planning and scheduling.

Chapter 9 discusses the subject of inspections and testing as part of the computer validation process.

Chapter 10 covers the subject of computer systems qualification, including sample qualification using the GAMP software categories.

Chapter 11 discusses the SLC documentation and its relationship with computer systems validation.

Chapter 12 covers the issues of procedural controls in regulated operations.

Change management and training are covered in Chapters 13 and 14, respectively.

Chapter 15 discusses security, the backbone of Part 11. This chapter introduces the key security services, user or data authentication, and access control.

Chapter 16 covers the issue of Source Code.

Chapter 17 covers supplier qualifications.

Chapter 18 shows how to maintain the state of the validation in computer systems. After the system has been released for operation, the maintenance activities take over.

Chapter 19 briefly discusses Part 11 remediation activities.

Chapter 20 covers the important subject of Operational Systems Checks.

Chapter 21 covers the most recent Compliance Policy Guide 7153.17 applicable to computer systems. The scope of this guide is Part 11.

Chapter 22 addresses specific issues on electronic records.

Chapter 23 covers the subject of electronic signatures and their implementation based on the requirements of Part 11.

Chapter 24 discusses how hashing, encryption, and digital signature technologies can be used to support Part 11.

Chapter 25 guides the reader on how to put together all the elements discussed in this book.

Chapter 26 introduces the reader to the future of software engineering and suggests how QA and QE must adapt to these changes.

Chapter 2

Validation Overview

WHAT IS A COMPUTER SYSTEM?

The term 'computer system' can define any of the following: desktop systems; client or server systems; automated process control and laboratory systems; host-based systems; data acquisition and analysis systems; and all associated software. The associated software comprises application software or firmware, system software, and computer system supporting documentation.

In the regulatory context, computer systems are integrated into the operating environment (Figure 2–1[1]). The operating environment

> **Regulatory guidance**
>
> *Computer systems shall be validated. The computer validation must ensure accuracy, reliability, consistent intended performance, and the ability to discern invalid or altered records.*
>
> Department of Health and Human Services, Food and Drug Administration, *21 CFR Part 11, Electronic Records; Electronic Signatures,* Federation Register 62 (54), 13430–13466, March 20, 1997.

may include the process or operation being controlled or monitored by the computer system, the procedural controls, process-related documentation, and the people. Computer systems performing regulated operations may either control the quality of a product during its development, testing, manufacturing, and handling processes; manage information business operations; manage data used to prove the safety; efficacy and quality of the product and formulation; and provide data for drug submissions.

WHAT IS A COMPUTER SYSTEMS VALIDATION?

The validation of computer systems performing regulated operations provides

confirmation by examination and provision of objective evidence that computer system specifications conform to user needs and intended uses, and that all requirements can be consistently fulfilled.[2]

It encompasses:

---

[1] R.R. Herr and M.L. Wyrick, A Globally Harmonized Glossary of Terms for Communicating Computer Validation Key Practices, *PDA Journal of Pharmaceutical Science and Technology*, March/April 1999.

[2] FDA, draft of *Guidance for the Industry: 21 CFR Part 11; Electronic Records; Electronic Signatures: Glossary of Terms*, August 2001.

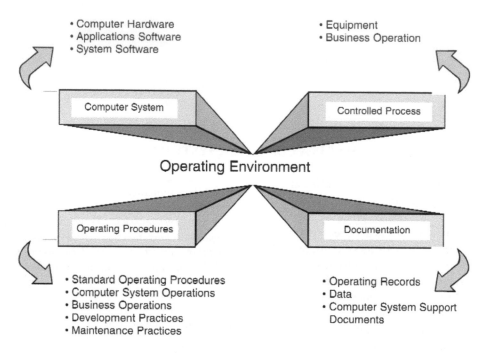

- Computer Hardware
- Applications Software
- System Software

- Equipment
- Business Operation

Computer System

Controlled Process

Operating Environment

Operating Procedures

Documentation

- Standard Operating Procedures
- Computer System Operations
- Business Operations
- Development Practices
- Maintenance Practices

- Operating Records
- Data
- Computer System Support Documents

Figure 2–1. A Computer System and the Operating Environment.

planning, verification, testing, traceability, configuration management, and many other aspects of good software engineering ... that together help to support a final conclusion that software is validated.[3]

The key words in the preceding paragraph are 'objective evidence.'

Objective evidence is generated following completion of a number of formal and informal activities, many of which must be completed in a predefined order. These activities comprise the SLC. The work products of each phase in the SLC provide the objective evidence that is required to demonstrate that the computer system conforms to the needs and intended uses of the user, and that all requirements can be consistently fulfilled.

WHY DO WE VALIDATE COMPUTER SYSTEMS?

Computer systems are validated for two important reasons: to ensure that good business practices are followed, and to satisfy regulatory agency requirements.

The main business reasons to validate computer systems are to demonstrate conformance with the system requirements specification, to increase acceptance of the systems by end-users, and to avoid high maintenance costs. Regarding maintenance costs, Figure 2–2 depicts the relative cost in 1976 of repairing software. At present, the proportions are similar. The longer a defect is uncovered, the more expensive it is to repair.

---

[3]   FDA CDRH, *General Principles of Software Validation; Final Guidance for Industry and FDA Staff*, January 2002.

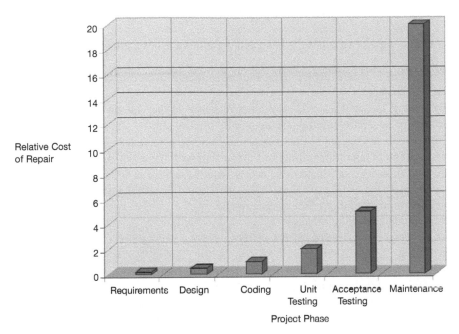

Figure 2–2. Relative Cost of Repairing Software Error.[4]

Taking as an example the conventional waterfall development methodology, an error introduced during the technical design activity will cost twice as much to fix during the program build stage. Each error uncovered during the operational life period will cost 20 times more to fix it during the coding stage.

Up to 50% of maintenance effort is spent fixing errors that have been present since the initial phase of the development.

The effect that verification and testing activities have on financial resources is noticeable. A potential saving of 50% on software maintenance can make a sizeable impact on overall life cycle cost.

Good business practices include assurance of the trustworthiness of the data and information. Other reasons to satisfy good business practices include:

- demonstrating the suitability of computer hardware and software to perform the assigned task
- facilitating the relationships between the company and its suppliers
- improving the level of control and maintainability

The reasons for satisfying regulatory agency requirements include:

- minimizing regulatory actions
- maintaining a positive relationship with regulatory agencies
- expediting submissions to and approval by the FDA
- avoiding product recalls and negative publicity

---

[4]   B.W. Boehm, Software Engineering, *IEEE Transactions on Computers*, 1976.

KEY PROJECT ELEMENTS

Key elements that must be successfully executed on computer implementation projects are:

- selection of a development, deployment, and/or maintenance methodology that best suits the nature of the project
- establishment of intended use and proper performance of the system
  - selection of hardware based on capacity and functionality
  - identification and testing of the operational limits to establish production procedures
  - identification of operational functions associated with the users, sequencing, regulatory, security, company (e.g., standards), safety requirements, and so on
  - identification and testing of 'worst-case' operational or production conditions
- accuracy and reproducibility of the testing results
- project documentation requirements[5]
  - written design specification[6] that describes what the software is intended to do and how it will do it
  - a written qualification/test plan based on the design specification,[7] including both structural and functional analysis, as applicable
  - test results and an evaluation of how these results demonstrate that the predetermined design specification has been met (e.g., requirements traceability analysis)
- availability of procedural controls to maintain the operating environment of the system environments
- any modification to a component of the system and its operating environment must be evaluated to determine the impact on the system. If necessary qualification or validation will have to be re-executed, partially or totally

WHICH SYSTEMS SHOULD BE VALIDATED?

Computer systems performing regulated operations shall be validated. Typical examples of such systems are:

- systems that control the quality of product during its development, manufacturing, testing, and handling processes
- systems that create, modify, maintain, archive, retrieve, or transmit data used to prove the safety, efficacy, and quality of product and formulations
- systems that provide information on which decisions are made affecting fitness for purpose of the product
- systems that create, modify, maintain, archive, retrieve, or transmit records that must be available for inspection by a regulatory agency
- systems that provide data or reports that are sent to regulatory agencies

The depth and scope of the validation depends on the category of the software, and the

---

[5] FDA, *Guidance for the Industry: Computerized Systems Used in Clinical Trials*, April 1999.
[6] In this context a design specification may include a set of specifications containing the project requirements, technical, configuration, functional, and/or technical design.
[7] FDA, *Guidance for the Industry: Computerized Systems Used in Clinical Trials*, April 1999.

complexity and criticality of the application. Refer to Chapter 7 for details of the computer system validation process.

INTRODUCTION TO THE COMPUTER SYSTEMS VALIDATION PROCESS

The system life cycle (SLC) is the *'period of time that begins when a product is conceived and ends when the product is no longer available for use.'*[8]

The development model associated with the SLC contains the software engineering tasks and associated work products necessary to support the computer system validation effort. It breaks the systems development process down into sub-periods during which discrete work products are developed. This approach leads to well-documented systems that are easier to test and maintain, and for which an organization can have confidence that the system's functions will be fulfilled with a minimum of unforeseen problems.

The development model contains specific inspection and testing tasks that are appropriate for the intended use of the computer system.

The naming of each sub-period depends on the development model used. The following enumerates development methodologies common in the FDA-regulated industry.

- Classical Life Cycle – System Engineering, Analysis, Design, Code, Testing and Maintenance
- PDA/PhRMA's Computer System Validation Committee – System Definition, Design, Unit Testing, Integration Computerized System, System Module Integration, System (Total) Integration, Operation, and Maintenance
- IEEE – Concept, Requirements, Design, Implementation, Test, Installation and Checkout, Operation and Maintenance
- FDA[9] – Requirements, Design, Implementation, Test, Installation and Checkout, Operation and Maintenance
- CDRH[10] – Specification/Requirements, Design, Implementation, Verification and Validation, Maintenance
- FDA/NCTR[11] – Initiation, Requirements Analysis, Design, Programming and Testing, System Integration and Testing, System Validation, System Release, Operation and Maintenance
- EU Annex 11 – Planning, Specification, Programming, Testing, Commissioning, Documentation, Operation, Monitoring and Modifying
- DOD–STD–2167A – System Requirements Analysis/Design, Software Requirements Analysis, Preliminary Design, Detailed Design, Coding and Computer System Unit Testing, Computer Software Component Integration and Testing, Computer Software Configuration Item Testing, System Integration and Testing

---

[8] ANSI/IEEE Std 610.12–1990, *Standard Glossary of Software Engineering Terminology,* Institute of Electrical and Electronic Engineers, New York, 1990.

[9] FDA, *Software Development Activities*, Reference Materials and Training Aids for Investigation, July 1987.

[10] FDA, *Review Guidance for Computer Controlled Medical Devices Undergoing 510(k) Review*, Office of Device Evaluation Center for Devices and Radiological Health, 1991.

[11] Office of Regulatory Affairs of the Food and Drug Administration and the National Center for Toxicological Research, *Computerized Data Systems for Nonclinical Safety Assessment*, October 11, 1987.

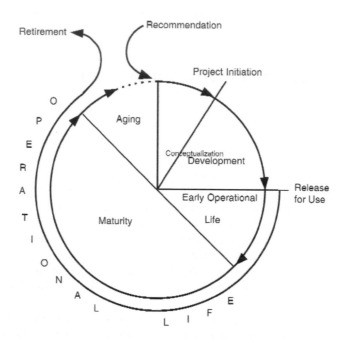

Figure 2–3. System Life Cycle focusing on Software Engineering Key Practices.

- Fourth Generation Techniques – Requirements gathering, 'Design' strategy, Implementation using 4GL, Product
- Object Oriented – The object-oriented (OO) model stresses the notion of software component reuse and comprises: requirements gathering, analysis, design, implementation (browse library or creation of reusable components), prototype build, and testing. The creation of reusable components is a development model within the OO model, and comprises the following: specification, design, code, unit test, integration into the prototype, and integration testing

Figure 2–3[12] depicts an SLC adapted to different system acquisition strategies and software development models. It is focused on software engineering key practices. It does not specify or discourage the use of any particular software development method. The acquirer determines which of the activities outlined by the standard will be conducted, and the developer is responsible for selecting methods that support the achievement of contract requirements. A modifiable framework must be tailored to the unique characteristics of each project. It includes the following periods:

- conceptualization
- development

---

[12] R.R. Herr and M.L. Wyrick, A Globally Harmonized Glossary of Terms for Communicating Computer Validation Key Practices, *PDA Journal of Pharmaceutical Science and Technology*, March/April 1999.

- early operational life
- maturity
- aging

Project Recommendation, Project Initiation, Release for Use, and Retirement are 'events' that must be considered as phase gates or major decision points, and that include formal approvals before development can proceed to the next period.

Certain discrete work products are expected when evidencing the development and maintenance work of computer systems compliance to regulatory requirements. The selected SLC should specify the overall periods and the associated events, and dictate the minimum requirements regardless of the chosen development method. These development methods and deliverables are also known as the development methodology.

The development methodology establishes detailed discrete work products, by phases and events, and by associated activities. The computer systems development approach must be consistent with the selected SLC; the most common development methodologies are the Waterfall Model, the Incremental Development Model, the Evolutionary Model, the Object Oriented approach, and the Spiral Model. Sample SLC activities and work products are detailed in Appendix E.

A critical function of the validation process is to provide an assurance that the development and operational life methodologies are consistent and followed carefully. The SLC and its associated development and operational life methodologies, applicable to computer systems performing regulated operations, shall be described in procedural control(s). The project team may have the authority to select a developmental and operational life methodology best suited to the nature of the system under development, and operational life methodologies that differ from those originally included in the related procedural control. In this case, the selected development or operational life methodologies must be clearly and satisfactorily explained in the validation plan or equivalent document.

It is the objective of FDA regulated companies in general to select the appropriate SLC and associated methodologies. Development and maintenance teams must receive adequate training in the use of the chosen methodology.

To evaluate adherence to the selected methodology, quality checkpoints (e.g., audits) are conducted during the project. Considering Figure 2–3 as an example, each event associated when the SLC is a checkpoint when quality checks can be conducted.

Detailed descriptions of the infrastructure for supporting computer systems validation can be found in Chapters 5 and 6. A sample computer systems development and maintenance methodology can be found in Chapter 7.

COMPUTER SYSTEMS VALIDATION FOR LOW-CRITICALITY AND/OR LOW COMPLEXITY PROJECTS

The methods contained in this book apply to any type of computer systems validation project. However, for low criticality and/or low complexity projects, certain steps in the computer systems project can be abbreviated: documentation may be shortened and combined; reviews can be less formal, but not less thorough. Appendix D, Sample Criticality and Complexity Assessment, provides guidelines for the management of and approach to low-criticality or low-complexity projects.

Chapter 3

USA Regulatory Requirements for Computer Systems[1]

The FDA's authority to regulate computer system functions in a regulated operation is derived from the FD&C Act.[2] Resulting from the FD&C Act, three (3) main items were influential in computer systems practices in the regulated operations: applicable predicate rule, CPGs, and Part 11.

In the predicate rules arena, good manufacturing practices (GMPs) applicable to the manufacture of pharmaceuticals and medical devices are the most crucial regulations to computer system practices in the regulated operations. GMPs are the specific regulations describing the methods, equipment, facilities, and controls required for producing human and veterinary products, medical devices, and processed food. The regulatory authority enforcing the GMPs is the FDA.

In the United States, drug GMPs were formally introduced in 1963 and included the regulation of computer systems. 21 CFR Part 211.2(b) emphasized computer backups and documentation, including keeping hardcopy of master formulas, specifications, test records, master production and control records, batch production records (batch production and control records), and calculations.

By 1978[3] the regulations combined 21 CFR Part 211.2(b) and 211.68. The product of this combination was the updated 21 CFR Part 211.68 (Automatic, mechanical, and electronic equipment).

In summary, 21 CFR Part 211.68 requires that:

- There must be a written program detailing the maintenance of the computer system
- There must be a system to control changes to the computer hardware and software, including documentation
- There must be documented checks of I/Os for accuracy. In practice, it is implied that all computer systems under the regulations must be qualified/validated
- There must be programs to ensure the accuracy and security of computer inputs, outputs, and data
- Computer electronic records must be controlled, and this includes record backup, security, and retention

Computer systems validation, implied in 21 CFR Part 211.68, established in 21 CFR Part 11.10(a), and defined in the recent draft FDA guideline,[4] is one of the most important

---

[1] López, O., FDA Regulations of Computer Systems in Drug Manufacturing – 13 Years Later, *Pharmaceutical Engineering*, Vol. 21, No. 3, May/June 2001.

[2] Section 510(a) (2)(B) of the Federal Food, Drug, and Cosmetic Act.

[3] FR, Vol. 43, No. 190, September 29, 1978.

[4] FDA, draft of *Guidance for Industry: 21 CFR Part 11; Electronic Records; Electronic Signatures Glossary of Terms*, August 2001.

13

requirements applicable to computer systems performing regulated operations. Computer systems validation is the confirmation (by examination and the provision of objective evidence) that computer system specifications conform to user needs and intended uses, and that all requirements can be consistently fulfilled. It involves establishing that the computer system conforms to the user, regulatory, safety, and intended functional requirements.

Computer systems validation is an element of the SLC. In addition to the software and hardware testing, other verification activities include code walkthroughs, dynamic analysis, and trace analysis.

To clarify the intent of 21 CFR Part 211.68, CPG[5] 7132a.07, 'I/O Checking' was published in 1982. According to this CPG, computers I/Os must be tested for data accuracy as part of the computer system qualification and, after the qualification, as part of the computer system's ongoing verification program.

CPG 7132a.07 is based on the realistic anticipation that computer I/O errors can occur on validated systems. Computer components such as logic circuits, memory, microprocessors, or devices such as modems and displays, can fail after they have been tested, just like any other mechanical part. Other sources of computer systems I/O malfunctions are electromagnetic interferences, e.g., radio-frequency interference, electrostatic discharge, and power disturbance. Software errors undetected during the validation process may also be a source of I/O errors. In order to detect errors before a computer system makes decisions using tainted data, an on-going monitoring program needs to be established and followed to verify hardware and software I/Os during the operation of the system.

The level, frequency, and extent of the I/O checking was suggested in the Federal Register of January 20, 1995 (60 FR 4091). The level and frequency of the I/O verifications must be guided by written procedure, and based on the complexity and reliability of the computer system.

Note: A CPG as described in 21 CFR 10.85 is considered an Advisory Opinion directed to FDA inspectors. These guidelines are the mechanisms utilized by the FDA to spread policy statements within the Agency and to the public.

Another topic relevant to computer systems in GMP environments is the management of application source code. Before 1985, 80% of computer systems were custom-built, making the source code a deliverable. Since then, software developers have begun to make off-the-shelf applications available: today, 80% of applications utilized to supervise manufacturing processes are configurable software. Many of the custom-built programs can be found, for example, in PLC. For configurable applications, the source code is the proprietary information of the developer, so the likelihood of acquiring the source code is very low.

The regulatory requirements applicable to source code can be found in CPG 7132a.15, 'Source Code for Process Control Application Programs,' and include consideration of the source code as master production and control records. Accordingly, those sections in the regulations relevant to master production and control records are applied to the computer application as source code. For custom-built applications, the program listings are considered to be source code; for configurable applications, the configurable elements or scripts are considered to be source code; and for off-the-shelf applications, the critical algorithms, parameters, and macro listings are considered to be source code.

CPG 7132a.08, 'Identification of 'Persons' on Batch Production and Control Records' issued in 1982, allows drug manufacturers to replace certain functions performed by operators with computer systems. Part 211.101(d) requires verification by a second person for components added to a batch. A single automated check is acceptable if it provides at least the same

---

[5] López , O., A Guide to Computer Compliance Guides, *Journal of cGMP Compliance*, January 1997.

assurance of freedom from errors as a double check. If it does provide the same assurance, the process does not gain by applying a redundant second check, which adds nothing to the assurance of product quality. The equivalency of an automated single-check system to a manual check must be shown, however, and this might not always be possible.

A major compliance issue is the administration of electronic records. Regulatory authorities hold owners of records required by existing regulation responsible for assuring the compliance of the recording and management of the electronic records. The FDA establishes such responsibility in the CPG 7132a.12, 'Vendor Responsibility' and Part 11 has now established the detailed requirements for such records.

In 1983, the FDA provided another important guideline applicable to computer hardware and software performing functions covered by the regulations. Their 'Guide to Inspection of Computerized Systems in Drug Processing' addresses the applicability of the regulations to computer systems. According to this guideline, computer systems hardware and software are considered equipment and records, respectively, within the context of the regulations. One year later CPG 7132a.11,[6] was issued, confirming the applicability of the regulations to computer hardware and software. In the absence of explicit regulations addressing computer systems, the regulations provide the implicit guidelines necessary to meet the agency's expectations. In accordance with CPG 7132a.11, Table 3–1 lists the primary sections in the regulations applicable to computer systems performing functions covered by the GMPs. Equivalent sections can be found in the other FDA regulations.

Table 3–1. US Drugs GMP.

US Drugs GMP	Description
211.22	Responsibilities of QC Unit
211.25	Personnel qualifications
211.42	Design and construction
211.63	Equipment design, size, and location
211.67	Cleaning and maintenance
211.68	Maintenance and calibration
211.68	Written procedures
211.68(b)	Record controls
211.68(b)	Validation of computer systems (implicit requirement)
211.100	Written procedures, deviations
211.101(d)	Double check on computer
211.105(b)	Equipment identification
211.180	General (records and reports)
211.180(a)	Records retention
211.180(c)	Storage and record access
211.180(d)	Records medium
211.182	Use of log(s)
211.188(a)	Reproduction accuracy
211.188(b)	Documentation and operational checks
211.189(e)	Records review
211.192	QC record review
211.220(a)[7]	Validation of computer systems (explicit requirement)

---

[6] FDA, CPG 7132a.11, *Computerized Drug Processing, CGMP Applicability to Hardware and Software*, 9/4/87.

[7] 1996 CGMP proposed regulations, FR, Vol. 61, No. 87, May 3, 1996.

Table 3–2. Reconciliation GMPs, EU Annex 11 and Part 11.

US Drugs GMP/CPG	EU GMPs[8]	Description	Part 11
CPG 7132a.07 CPG 7132a.08 211.68(b) 211.220(a)*	Annex 11-2	Validation	11.10(a)
211.180(a)	Annex 11-12, 11-13	Generation of accurate and complete copies of records	11.10(b)
211.68(b)	Annex 11-13, 14, 15, 16	Protection of records	11.10(c)
211.68(b)**	Annex 11-8	Limiting access to authorized individuals	11.10(d)
Comment paragraph 186, 1978 CGMP revision	Annex 11-10	Use of audit trails	11.10(e)
CPG 7132a.15 CPG 7132a.08	Annex 11-6	Use of operational system checks to enforce permitted sequencing of steps and events, as appropriate	11.10(f)
211.100; 211.188(b)	Annex 11-10	Authority checks	11.10(g)
211.68(b) n/a	Annex 11-8	Use of device (e.g., terminal) checks to determine, as appropriate, the validity of the source of data input or operational instruction	11.10(h)
211.25	Annex 11-1	CVs and training records for those using e-records	11.10(i)
211.180	Annex 11-4	Controls over systems documentation	11.10(k)
n/a***	Annex 11-8	Controls for identification codes/password	11.300
Subpart D	Annex 11-3, 4, 6, 13, 15	Controls for computer hardware	11.10(d)
211.180(a)	Annex 11-14	Record retention	Not covered

8   Wyn, Sion, Regulatory Requirements for Computerized Systems in Pharmaceutical Manufacture, *Software Engineering Journal*, Vol. 11, No. 1, 1996, pp. 88–94.

*   Proposed changes to CGMP, May 1996.

**   Sec. 211.68(b) requires appropriate controls over computer or related systems to ensure that only authorized personnel make changes in master production and control records or other records.

*** Implicit regulations can be found in 211.68.

The introduction in 1997 of Part 11 provided the formal codification applicable to computer systems performing FDA-regulated operations. One of the fundamental principles of Part 11 is that it requires organizations to store regulated electronic data in its electronic form, once a record has been saved to durable media, rather than keeping paper-based printouts of the data on file, as had been the long-term practice in organizations performing regulated operations[9]. If information is not recorded to durable media, the stored data will be lost and cannot be retrieved for future use.

When 'retrievability' is an attribute, the procedural and technological controls contained in Part 11 are essential to ensure record integrity. Only regulated electronic records that meet Part 11 can be used to satisfy a predicate rule.

The current federal legislation embodied in the Electronic Signatures in Global and National Commerce Act, defines electronic records in more general terms than Part 11. The primary requirement of the current federal legislation is whether data is accessible once it is put into storage, rather than the technology used. This definition considers data in transient memory as an electronic record. Therefore, one key element to be analyzed by both the FDA and the industry are the regulatory requirements for data that stored in transient memory, including any audit trail information.

The controls on records contained in 21 CFR 11 Subpart B reconcile earlier GMPs and policies. Table 3–2 associates the clauses in Subpart B 21 CFR Part 11 with the drugs GMP regulations, applicable FDA CPGs, and the EU GMP regulations.

CPG 7153.17 is the most recent CPG applicable to computer systems. Published in 1999, the objective of this CPG is to establish the criteria that the FDA will take into account when regulatory action is needed for a computer system found to be noncompliant with Part 11. Refer to Chapter 21.

Another key concept in the FDA regulatory environments is the security of computer resources. Chapter 14 covers this subject.

Regulatory authorities hold the owner of the records known to be required by existing regulation responsible for assuring the compliance of the computer systems that record and manage such records. The FDA defined this responsibility in CPG 7132a.12, 'Vendor Responsibility.'

MEDICAL DEVICES SOFTWARE

The primary FDA regulations applicable to medical device product software are 21 CFR 820.70(i), 21 CFR 820.30 and all of the CPGs discussed in this chapter. Validation requirements apply to software used as components in medical devices, to software that is itself a medical device, and to software used in production of the device or in implementation of the device manufacturer's quality system. According to 21 CFR 820.70(i) the automated system must be qualified for its intended use to ensure system performance. There must be an approved procedure and qualification protocol(s) in order to ensure that qualification is properly performed. The FDA also requires that electronic components of medical devices be regulated. An evaluation of component suppliers and qualification of device manufacturing line(s) are elements of this qualification.

---

[9] Note that Part 820 requires that 'results' of acceptance activities be recorded but not necessarily all raw data. 'Results' must have audit trails. Be sensitive to need for raw data during failure investigations under CAPA. Refer to Part 820 preamble, pp. 52631 and 52646.

In order to meet the quality system regulations, electronic components are considered to be similar to process equipment and components, and must be traceable from the suppliers to the user. According to the FDA, software development is primarily a design process. 21 CFR 820.30 contains the design control regulations applicable to medical devices. The application of this section to product software[10] includes all design activities necessary to obtain, review, implement, and validate a design specification.

The design specification describes, in a narrative and/or pictorial form, how the software will accomplish the software requirements and 'the interactions with the hardware to accomplish various functions of the device's design.'[11] The most recent FDA validation guidance,[12] which is exclusively applicable to medical devices, provides comprehensive information on the implementation of applicable software validation regulations to medical device computer systems performing regulated operations.

In the context of Part 11, one of the differences between Parts 211 and Part 820 is the scope of the electronic records that must be maintained. Part 21 requires the maintenance of electronic regulated data including raw data. Part 820 requires that the results of acceptance activities are recorded, but not necessarily all of the raw regulated data. The exception in 820 is that raw data is required during failure investigations. As in Part 211, results under Part 820 must have audit trails. Refer to Part 820 preamble, pp. 52631 and 52646.

THE FOOD INDUSTRY

FDA's authority to regulate the use of computers in food plants is derived from FD&C Act Section 402 (a) (3). Specifically, conventional food is governed by the CGMP in Manufacturing, Packaging, or Holding Human Food (21 CFR Part 110, April 2001). 21 CFR 110.40(d) provides the regulatory requirements applicable to computer systems:

Holding, conveying, and manufacturing systems, including gravimetric, pneumatic, closed, and automated systems, shall be of a design and construction that enables them to be maintained in an appropriate sanitary condition.

There is no additional explicit citation to computer systems in Part 110. Taking into account CPG 7132a.11, computer systems are referenced implicitly in multiple areas. Part 11 covers electronic record requirements in Part 110. FDA regulates dietary supplements under a different set of regulations than those covering 'conventional' foods and drug products (prescription and over-the-counter). Under the Dietary Supplement Health and Education Act of 1994 (DSHEA), the dietary supplement manufacturer is responsible for ensuring that a dietary supplement is safe before it is marketed. FDA is responsible for acting against any unsafe dietary supplement product after it reaches the market. At the present time, moves are being made to ensure that manufacturers of dietary supplements register with FDA, or get FDA approval, before selling such supplements. Manufacturers must make sure that product label information is truthful and

---

[10] O. López, Applying Design Controls to Software in the FDA-Regulated Environment, *Journal of cGMP Compliance*, July 1997.

[11] FDA, *Application of the Medical Device GMP to Computerized Devices and Manufacturing Processes*, May 1992

[12] FDA, *General Principles of Software Validation; Final Guidance for Industry and FDA Staff*, January 2001.

not misleading. FDA's post-marketing responsibilities include monitoring safety, e.g., voluntary dietary supplement adverse event reporting, and product information, such as labelling, claims, package inserts, and accompanying literature. The Federal Trade Commission regulates dietary supplement advertising. At this moment there is no regulation covering the manufacturing of dietary supplement. Currently there is a proposed ruling for the Current Good Manufacturing Practice in Manufacturing, Packaging, or Holding Dietary Supplements.[13] Under Equipment and Utensils, explicit reference on computer systems performing functions in FDA-regulated products:

(a) (6) Holding, conveying, and manufacturing systems, including gravimetric, pneumatic, closed, and automated systems, shall be of a design and construction that enables them to be maintained in an appropriate clean condition

indicates that computer-related regulatory provisions for the food industry would be equally applicable to the dietary supplements industry.

---

[13] FR, Vol. 62, No. 25, February 6, 1997.

Chapter 4

New Computer Systems Validation Model[1]

In the past, the approach to computer qualification or validation and associated configuration management in the regulated industries was based on key practices. The foundations of these key practices are contained on publications such as:

* *Guideline on General Principles of Process Validation*, May 1978
* *Compliance Policy Guidelines*, September 1982
* *Guide to Inspection of Computerized Systems in Drug Processing*, February 1983
* *PMA Staying Current Series* (series started in May 1986)
* *Application of the Medical Device GMPs to Computerized Devices and Manufacturing Processes*, November 1990
* *Good Automated Manufacturing Practices (GAMP)*, February 1994
* PDA's *Validation of Computer Related Systems*, October 1994
* *General Principles of Software Validation; Final Guidance for Industry and FDA Staff*, June 2001
* PDA's A Globally Harmonized Glossary of Terms for Communicating Computer Validation Key Practices, *PDA Journal of Pharmaceutical Science and Technology*, Vol. 53, No. 2, March/April 1999
* *Best Practices for Computerized Systems in Regulated GxP Environments*, PIC, Draft Rev. 3.01

The constraint of key practices is that they 'need to be monitored and evaluated periodically to ensure that they are suitable, and to keep them current with industry and regulatory trends.'[2] Across the industry, key practices were not always implemented at the same time, causing divergence in the implementation of the related key practices.

Part 11 provides the explicit and current regulatory trends applicable to computer systems performing regulated operations. Now that the regulatory expectations are clearly established, the implementation of Part 11 is contingent on the availability of appropriate applications, which are compliant with the regulation. In the absence of technologies supporting Part 11, the required controls are procedural in nature.

Figure 4–1 depicts a 21 CFR Part 11 Model developed by the author of this book[1] helps determine Part 11 requirements or consideration during the development and qualification of computer systems based on three (3) main characteristics:

* whether the computer system is closed or open
* whether the system creates electronic records
* whether the system uses electronic signatures

---

[1] López , O., FDA Regulations of Computer Systems in Drug Manufacturing – 13 Years Later, *Pharmaceutical Engineering*, Vol. 21, No. 3, May/June 2001.

[2] Grigonis, Subak, and Wyrick, Validation Key Practices for Computer Systems Used in Regulated Operations, *Pharmaceutical Technology*, June 1997.

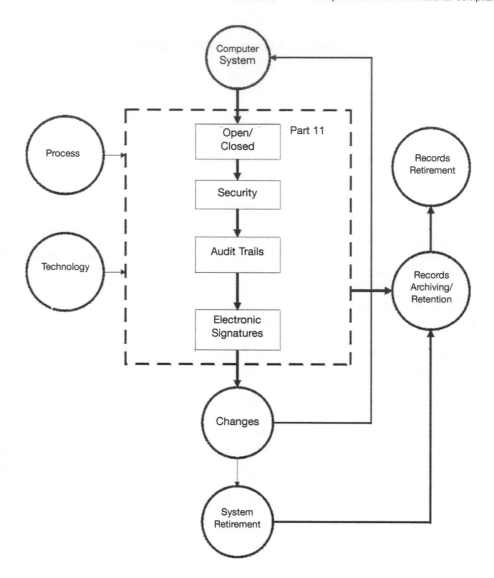

Figure 4–1. 21 CFR Part 11 Model.

In general, the elements to be verified and/or qualified during software implementation are

- the type of system (open or closed)
- security functions
- audit trails
- operations controlled by the computer systems (operational checks)
- implementation of the technology to support the process and Part 11

These two latter elements are not identical on all systems. Electronic signatures are tested when the technology is implemented. An element out of the scope of this model is the retention of electronic records, but the model should be used to verify and validate the implementation of the system(s) that will hold these records. For applications using electronic signatures, the current validation practices requires inspections and/or testing for many of the following technical controls:

1. **Open/Closed systems**
2. **Security**
    a. System security
    b. Electronic signature security
    c. Identification code and password maintenance
        - Identification code and password security
        - Passwords assignment
    d. Document controls
    e. Authority, operational, and location checks
    f. Records protection
3. **Operational Checks**

4. **Audit Trails**
    a. Audit mechanism
    b. Metadata
    c. Display and reporting
5. **Electronic signatures**
    a. E-sign without biometric/behavioral identification
    b. E-sign with biometric/ behavioral identification
    c. Signature manifestation
    d. Signature purpose
    e. Signature binding
6. **Certification to FDA**

A subset of the above is applicable for hybrid computer systems.[3] The implementation of these requirements has been discussed in other articles.[4,5]

Figure 4–2 depicts the progression of the computer systems validation practices, from the current best practices to the 21 CFR Part 11 Model. The transition program consists of educating the regulated industry, assessing current computer systems, and implementing the regulations using the appropriate technologies.

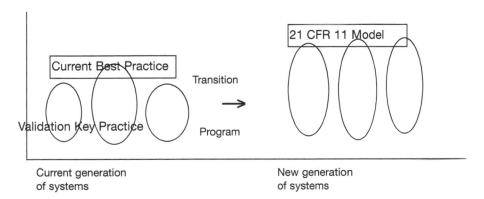

Figure 4–2. Progression of Computer Systems Validation Key Practices.[6]

---

[3] In hybrid systems, some portions of a record are paper and some electronic.

[4] López, O, Automated Process Control Systems Verification and Validation, *Pharmaceutical Technology*, September 1997.

[5] López, O, Implementing Software Applications Compliant with 21 CFR Part 11, *Pharmaceutical Technology*, March 2000.

[6] *A Partnership Approach to Achieving Regulatory Compliance for Electronic Records and Signatures*, paper presented at the October 1999 IMPACC Conference.

This model can be easily applied to the new generation of computer systems by incorporating the requirements contained in Part 11 at the beginning the development process. For the current generation of systems, an assessment will have to performed to evaluate the level of conformity with the regulation.

Examples of the applicability of this model can be found in Appendix C.

Chapter 5

Computer Validation Management Cycle

The implementation or update of a computer system is planned and executed in accordance with the organization's own project management practices. It is the project manager's responsibility to take into account the technological requirements, standards, procedural controls, regulatory requirements and related guidelines, and industry standards.

During the implementation or update process, the introduction of a computer validation management cycle will provide a top-down approach to integrate computer systems validation strategies in order to support the project. Figure 5–1 suggests a computer validation management cycle.

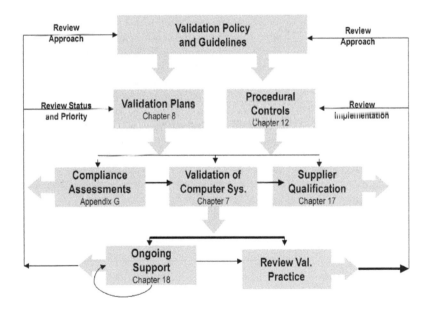

Figure 5–1. Computer Validation Management Cycle.

As with any process, an improvement mechanism is required in order to enhance the computer validation management cycle. The improvement mechanism will be identified based on experience in using the SLC, validation practices, and the analysis of software metrics. The following sections describe the overall scope of each document or activity. Detailed description of each activity can be found in other chapter of this book.

VALIDATION POLICIES[1]

Policies define the general principles and philosophy that are required within an organization. Each company should have written and approved validation policies that communicate the expectations of senior management for the execution of a validation project.

Policies are intended to establish responsibilities and expectations. All computer systems validation procedures, guidelines, and strategies must be consistent with the Validation Policy.

An example of a corporate-wide validation policy is contained in Sidebar 5–1.

In the context of Part 11, it is expected that information shall be characterized and information quality incorporated into the computer systems development and operational life periods.

Company name is committed to ensuring that the products, equipment, facilities, computerized systems, and support systems directly involved in the manufacture, testing, control, packaging, holding and distribution of marketed products are compliant with cGMP regulations. This commitment extends to the manufacturing, facilities, utilities, processing, cleaning, computerized systems, analytical and microbiological methods, packaging and support systems involved in the development and production of pharmaceutical products subject to cGMP regulations. To attain this level of compliance, management will ensure that proper validation and qualification will be carried out at all stages of the product development cycle.

Any new equipment, process, product, facility or system will be prospectively validated prior to becoming operational in the manufacture of finished dosage forms. Concurrent validation and qualification will be considered for any existing equipment, facility, product or system where limited documentation of appropriate technology background exists, and an acceptable level of confidence can be established by reviewing and documenting the operating history. Concurrent validation will be employed where applicable, to certify that an appropriate level of process control exists for all pertinent products, equipment, facilities and systems.

Retrospective validation or qualification is discouraged, but may be applicable in some situations, where a statistically significant number of batches with consistent formulas, procedures and analytical methods are available. Sufficiently detailed past processing and control records must be available for retrospective validation studies to be considered.

Sidebar 5–1. Sample Policy.

VALIDATION GUIDELINES

Validation guidelines must comply with all formal procedural controls established by the company/organization. They are internal documents providing detailed procedures, activities and sample products necessary to successfully execute a computer systems or related project.

Typically, a guideline provides *recommendations*, not *requirements*, and represents current thinking on the topic covered within the guideline. An alternative approach may be used, however, if such an approach satisfies the requirements of the applicable procedural controls

---

[1] PDA Committee on Validation of Computer-Related Systems, Technical Report No. 18 Validation of Computer-Related Systems, *PDA Journal of Pharmaceutical Science and Technology*, Vol. 49, No. S1, 1995.

and predicate regulation. It is often the practice of the FDA-regulated industries to not refer to guidelines in SLC related documentation. Guidelines and operating procedures may be contained in a Quality Plan, model specified in ISO 0005:1995, which defines and documents how the requirements for quality will be met for a project. Quality Plans and Validation Project Plans complement each other.

VALIDATION PLANS

Refer to Chapter 8.

PROCEDURAL CONTROLS

Refer to Chapter 12.

COMPLIANCE ASSESSMENTS

Compliance assessments for computer systems are performed periodically, based on the applicable predicate regulatory and Part 11 requirements. These assessments must be performed in order to identify any functional gaps, and/ or procedural gaps, which may be present for each computer system implemented. The analysis will determine if operational, maintenance or security controls, specific to the system, provide a controlled environment ensuring the integrity of the electronic records and/or signatures as stated in the regulatory requirements. Additional information can be found in Chapter 19.

VALIDATION OF COMPUTER SYSTEMS

Refer to Chapter 2.

SUPPLIER QUALIFICATION

The qualification of computer technology suppliers and contract developers is an ongoing activity. The objective of this process is to evaluate the computing environment and technology products of the supplier, against established quality standards, managing any noted deficiencies that are discovered, and working with the supplier to improve the quality of its products, services, and documentation. Chapter 17 provides information on the supplier qualification activities.

ONGOING SUPPORT SYSTEMS

After the system has been released for operation, system maintenance activities take over. The importance of such activities is characterized by recent FDA remarks related to the lack of change control management by regulated organizations. The FDA's analysis of 3140 medical device recalls conducted between 1992 and 1998 reveals that 242 of them (7.7%) are attributable to software failures. Of those software related recalls, 192 (or 79%) were caused by software defects introduced when changes were made to the software, *after its initial production and distribution.*

The maintenance activities must be governed by the same procedures followed during the Development Period.

Chapter 18 provides information on ongoing support systems.

Chapter 6

Computer Validation Program Organization

Many groups, both within and outside the organization, support the validation of computer systems. It is the responsibility of Executive Management to provide adequate resources to support the achievement of compliance in the computer systems validation area.

Responsibilities are often similar across multiple functional groups such as Information Technology, Engineering, Laboratories (QA/R&D), Clinical Research, Manufacturing/ Operations, contract developers, and computer technology suppliers.

In order to manage the execution and supervision of computer systems validation activities, there needs to be an organizational structure established. This chapter suggests an organizational structure for supporting to a computer validation program.

ORGANIZATIONAL MODEL

This model (Figure 6–1) is an example of how computer systems validation and its related activities can be organized and responsibilities allocated. Each organization should prepare a document to identify the roles and responsibilities appropriate for its business environment.

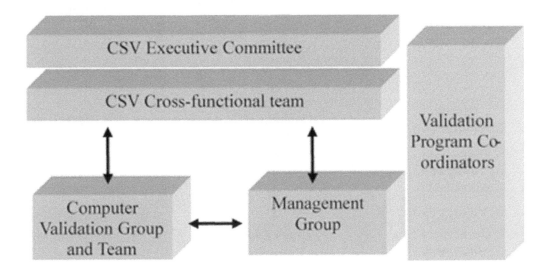

Figure 6–1. Organizational Model.

COMPUTER SYSTEMS VALIDATION EXECUTIVE COMMITTEE

The Computer Systems Validation (CSV) Executive Committee is composed of Validation, Compliance, Regulatory Affairs, and Information Technology senior management. Another possible member of this committee is a Process Innovator.

The CSV Executive Committee is responsible for:

- reviewing and approving computer systems validation guidelines
- reviewing and approving the computer systems validation master plans
- providing direction and guidance concerning the validation program
- arbitrating on organizational issues and validation program issues
- supporting the awareness and implementation of the CSV strategy through resourcing training, and procedure and guideline development

CSV CROSS-FUNCTIONAL TEAM

The Cross-Functional Computer Systems Validation Team is a multidisciplinary team (Figure 6–2) that represents the overall needs of the business. This team comprises representatives from research production, engineering, quality control, clinical research, quality assurance, and other groups as needed.

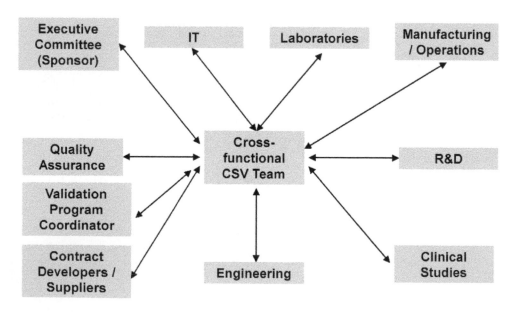

Figure 6–2. Cross-Functional Team.

The members and their deputies are assigned by the CSV Executive Committee. For *ad hoc* activities, the team can be complemented with representatives from departments performing activities that support computer validation. Examples of such departments are: Purchasing, IT,

QA Validation, QA Compliance, Operations, R&D, Engineering, and representatives from each manufacturing facility concerned.

The Cross-Functional CSV Team is responsible for:

- reviewing and approving the CSV procedural controls
- organizing the implementation of the validation policy, validation guidelines, and computer system validations procedural controls
- developing and maintaining the CSV approach to be followed by the validation team
- reviewing validation plans and protocols as needed
- assisting the validation program coordinator in compiling the validation master plans
- developing and maintaining a QA framework for CSV activities
- assisting QA Compliance during audit of a computer system
- support computer-related company and/or executive committee strategies

The importance of the QA unit in validation is evident from the regulations. The QA unit has the responsibility and authority for:

- reviewing and approving the validation policy, validation guidelines and validation procedural controls
- reviewing the computer systems validation work and documentation and signing to verify the correct execution of the validation plan, protocols, and summary report for computer systems performing regulated operations
- assessing the compliance of computerized systems to the validation policy, validation guidelines, validation procedural controls, and to related predicate regulations
- auditing the readiness of computer systems for a regulatory inspections
- supporting Line Management Groups during regulatory inspections
- documenting the results of audits in QA audit reports
- ensuring that any audit report issues are resolved
- organizing capability assessments on external and internal suppliers

CSV GROUPS AND TEAMS

The CSV Groups and Teams mainly focus on the implementation of computer systems validation projects.

The CSV Group is responsible for the delivery of a computer validated system. This includes:

- Performing the validation work for GxP computer systems from the plan through to the report and producing the required documentation
- Ensuring that the computer system is maintained in a validated state by ongoing qualification or validation of the platform

The member of the CSV Group include the:

- System Validation Team designated by the Management Group
- Supplier of the Standard/Configurable Software Packages used and/ or the System Integrator (software system and platform)
- Platform Owner

The CSV Team includes persons in the organization who plan and execute the CSV activities, including the System Owner, System Users, Validation Coordinator, and specialists in various fields related to computer systems validation. The CSV Team is responsible for:

- developing, approving and executing the CSV Plan
- developing and approving protocols and qualifications reports and
- writing the Validation Summary Report

One of the most important members of the CSV team is the Validation Project Coordinator. Responsible for planning, coordinating, and reporting the validation activities throughout a project, the validation coordinator may act as the liaison between the Validation Program Coordinator, other teams and groups in the organization. The Validation Project Coordinator will act as a consultant and must be familiar with the contents of the validation practices and procedures; and must have participated in appropriate training activities in the area of computerized systems validation.

When embarking on the implementation of a regulatory computerized system, it is recommended that the project managers appoint an individual to act as the validation project coordinator. The involvement of the validation coordinator on a project will vary depending on the size of project. For a small project, the validation coordinator role may require only a part-time assignment. For a large project, the validation coordinator role may require the assignment of a full-time person.

THE MANAGEMENT GROUP

The Management Group consists of the:

- System Owner
- System User

The Management Group is responsible for:

- preparing, approving, and implementing procedural controls that are appropriate for the system
- managing the system throughout its entire project life cycle
- preparing the information needed by the Validation Project Coordinator for the computer validation master plan
- setting up, managing, and resourcing the CSV Team
- approving the Validation Reports for the computerized systems
- archiving the system validation documentation
- ensuring the validation status is maintained throughout the operational life of the systems
- providing support during regulatory inspection

VALIDATION PROGRAM COORDINATORS

This is a dedicated corporate consulting group. The principal task of this group is to provide consultation and to approve validation documentation; a benefit of this approach is that a consistent computer validation program is implemented across all sites and units. For any validation activity to be effective, an independent and qualified third party (someone who is neither a developer nor a user of the system) must review it. Although the developers and users of the system may generate most of the SLC products, a qualified third party should review the computer systems validation records and give objective assurance to management that the computer systems validation was properly carried out. This third-party review function can be performed by a dedicated validation group, with sufficient objectivity and expertise in computer systems testing and validation methods. Members of a CSV group should include system, software, and hardware engineers with experience in regulations and project management; a dedicated CSV group needs experience and knowledge of:

- computing infrastructures
- security and controls
- corporate, industry, and regulatory standards
- solving computer systems problems
- the roles and responsibilities of the integrators

The Validation Programs Coordinators may be part of the QA unit, Technical Services unit, or the Information Technology unit.

The advantages of the dedicated Validation Program Coordinators are that employees in such a group are totally dedicated to (and responsible for) the computer validation effort. Interaction with production and quality assurance scheduling is important and must be taken into consideration.

The value added by a dedicated CSV group to the validation process includes:

- reducing the learning curve to the Management Group, System Validation Team, and System Validation Group by providing validation consultation
- producing a smoother path for reviewing the manufacturing systems validation documentation and reviewing protocols at the start of the development effort
- reducing inefficiencies by ensuring the quality and consistency of the validation methodology
- improving working inefficiencies by reusing validation experience between similar computer systems and in manufacturing facilities

To summarize, the Validation Program Coordinators are responsible for:

- managing the CSV program, consisting of the Validation Policy, validation guidelines, validation procedural controls development and implementation, computer validation training, and other activities that have been identified
- coordinating the development of validation guidelines and validation procedures and organizing regular reviews, and the development and review of generic validation plans and protocols
- reviewing computer validation work and documentation and signing to verify the correct execution of the validation plan, protocols and summary report for computer systems and to ensure the uniformity and quality of the approach
- compiling the Computer Validation Master Plan
- keeping an inventory of the validation status of the computer systems
- providing support, consultancy, tools (e.g., templates) and expertise on CSV matters to the CSV teams
- supporting QA during audits that assess the readiness of computer systems for regulatory inspections

- supporting Line Management Groups during regulatory inspections
- developing and providing validation training based on procedural controls

Chapter 7

The Computer Systems Validation Process

It is an FDA requirement that regulated organizations ensure that *all* computer systems performing functions under FDA predicate regulations conform to the requirements contained in Chapter 3. In the FDA regulation context, computer systems validation is an element of the SLC. The scope includes all the practices and tools necessary to manage the development and maintenance of the deliverable system.

All new computer systems must be **prospectively validated** before going into production.

For existing computer systems, **concurrent validation** may be considered where limited documentation on the appropriate technology exists, and where an acceptable level of confidence can be established by reviewing and documenting the operating history.

For existing computer systems, **retrospective evaluation** is discouraged. It is extremely difficult to evaluate a computer system retrospectively, being generally more costly and time consuming than prospective validation. Retrospective evaluation should be used only as a corrective measure in response to deficiencies noted concerning prior validation efforts. See Chapter 19 for a brief discussion on this subject.

Any deviations raised during the application of a computer system validation procedure shall be documented following the applicable deviation recording procedure.

The following describes the process for developing and maintaining computer systems in an FDA context.

The depth of validation and the steps of the validation process to be followed will vary depending on the category of software (e.g. as the one described in GAMP)[1] and the complexity and criticality of the system.

The depth of the required steps is estimated using the Criticality and Complexity Assessment and may be performed during the early Project Conceptualization stages of a project; and re-evaluated at Project Initiation. A sample Criticality and Complexity Assessment can be found in Appendix D.

Each phase of the SLC must be controlled to maximize the probability that a finished system meets all quality, regulatory, safety, and specification requirements. If an SLC approach is applied properly, no additional work will be required to validate a system. For each SLC period and event, computer systems validation requires that the development processes are documented work products. As explained in Chapter 2, phase gate verification activities performed during each event may be a perfect place to review and quantify the quality of all products needed to support the next phase.

The elements to be included as part of the computer systems validation process will vary depending on the category of the software. It is the practice in the FDA-regulated industry to take into account five (5) categories of software.

---

[1] *GAMP Guide for Validation of Automated System*, Version 4.0, ISPE (GAMP Forum), December 2001.

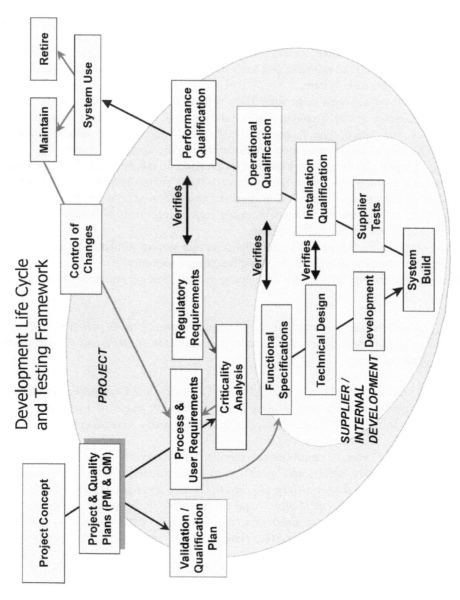

Figure 7–1. Conventional Waterfall Model.

[2] K. Baker and T. Simmons, *Software Management Basics* (ISBN 0–954979607), Sue Horwood Publishing Ltd., Storrington, U.K.

- Operating systems (e.g., Windows)
- Standard instruments (e.g., bar code readers, some laboratory instruments)
- Standard software packages (e.g., spreadsheets, databases)
- Configurable software packages, (e.g., LIMS, SCADA, building automation systems)
- Custom-built systems (e.g., automated process control systems)

Chapter 10 provides a comprehensive description of the validation elements to be considered for each software category.

Figure 2–3 illustrates a typical SLC. Using as an example, a high-criticality and high-complexity, custom-built system, the conventional 'V' model (Figure 7–1) as framework, and the typical waterfall methodology as development methodology, the following sections summarize the typical project activities.

Detailed descriptions by SLC Periods can be found in Appendix E.

- *1st Period — Conceptualization*
  During this period the computer system high-level functionality is agreed upon with the users, system owner, developers, and quality for the operation and automation of the operation. Part 11 considerations are a key element in this phase, which includes the following activities.

  - Initialization of the project to automate or to re-engineer an operation
    - Review project proposal
    - Develop operation study
    - Perform preliminary evaluation of the technology available
    - Refine scope of the system and confirm strategic objectives
    - Develop a process/operating description, associated restrictions, interfaces, proposed solution, and critical operational checks/sequencing, including how Part 11 will affect and interact with the process/operation
    - Approve the operational feasibility study
    - Perform a preliminary Criticality and Complexity analysis

  - Business and regulatory decision to continue or to hold the project

Appendix E contains a comprehensive description of the Conceptualization Period.

- *2nd Period — Development*
  The purpose of the Development Period is to specify, design, build, test, install, and support the business operation to be automated or updated, and approved in the Conceptualization Period using a structured process. Part 11 must be taken into account during all of these activities. The development process involves the following steps:

  - Requirements gathering
    Interviews, surveys, studies, prototypes, demonstrations, and analysis methods may be used to gather requirements

    - Gather the requirements for the systems including functional (e.g. operational checks) requirements, nonfunctional (e.g., coding standards) requirements, users, company-wide regulatory compliance (e.g., Part 11 technical control), safety, process, and other applicable requirements
    - Characterize information, assess its value to the organization, and incorporate information quality as part of the project plan

> Conduct a system (hardware, software, and process) risk analysis. New requirements may be found as the result of the risk analysis. Any new requirements must be documented in the requirements specification deliverable
> Data definition, database requirements, and records storage requirements
> Develop a requirements specification deliverable
> Review the requirements specification deliverable against the conceptual system requirements established during the Conceptualization Period. Any discrepancy must be resolved and the documentation amended
> Develop a validation plan
> Complete a risk analysis and a criticality/complexity analysis
> Complete change control and approvals

– System specification
The system specification deliverable describes in detail what the system should do, but not how (in terms of technology) it should be done. It is based on the requirement specification deliverable and the supplier's solution.

> Approved requirements specification deliverable
> If the software will not be developed in house, conduct a software supplier audit
> If elements of the hardware are unique, conduct a hardware supplier audit
> Select supplier(s)
> Gather all of the measurable operational parameters and limits for the system
> Develop high-level flow diagrams and general screen layouts
> Review the system specification deliverable. The objective of this review is to verify that the all of the specifications are correct, consistent, complete, unambiguous, testable, modifiable, and traceable to the concept phase documentation
> Perform and document a traceability analysis between the requirements specification deliverable and system specification deliverable
> Specify Part 11 technical controls
> Prepare and approve system specification deliverable
> Revisit the risk analysis and the criticality and complexity assessments based on the system specification deliverable
> Review validation plan based on the system specification deliverable
> Approved validation plan
> Begin to plan the test activities, including unit testing, integration testing, factory acceptance testing, site acceptance testing, and system qualifications

– Technical design
System requirements are applied to the software design. During the technical design phase, each requirement described in the system specification deliverable is implemented. Also, subsystem components and interfaces, data structures, design constraints, and algorithms are developed as part of the system design decomposition process. This activity is very critical for medical device companies. The design inputs are contained in the requirements for the computer system and the design output(s) can be provided as part of the design specification or the deliverable for the activity concerned[3].

---

[3] O. López, Applying Design Controls to Software in the FDA-Regulated Environment, *Journal of EGMP Compliance*, Vol. 1, No. 4, July 1997.

> Design according to SLC procedures
  •• computer hardware and software architecture
  •• data structures
  •• flow of information. This design phase forms the input for the development of the integration test and operational checks
  •• interfaces
> Assemble/collate the design deliverables
> Perform design reviews. Verify if the risks that were previously identified were mitigated as part of the solution presented in the design
> Finalize the test planning
> Design the Part 11 technical controls
> Approve the technical design specification deliverable(s)
> Conduct in-process audit activities associated with the technical design process
> Revisit the risk analysis
> Begin planning the development of procedural controls for those Part 11 requirements that are not covered by technology

- Program build
> Provide the technical design product(s) to the programmers
> Translate the design to a target program
> Develop programs based on the development (e.g., coding) standards, and build Part 11 technical controls into the system
> Review code integrity using static analysis, audit(s), inspection(s), review(s), and/or walkthrough(s)
> Conduct software unit testing. The machine executable form of the code is tested to uncover errors in function, logic, or implementation
> Integrate modules and perform integration testing
> Test the Part 11 technical controls

- Preparation of the qualification protocols[4]
> An IQ is required if:
  •• new software/hardware is being installed
  •• existing software/hardware has been modified or relocated
  •• existing software/hardware was never qualified
> An OQ is required if:
  •• operating parameters and sequences (e.g., operational checks) are being qualified for new software/hardware
  •• sequences have been modified
  •• new operating parameters are being established for existing software/hardware
  •• existing software/hardware was never qualified
    In some environments, an OQ is required to test software in testing environment.
> A PQ is required if:
  •• PQ is always required to test software in the production environment

---

[4] Author's note: While IQ/OQ/PQ terminology has served its purpose well and is one of many legitimate ways to organize dynamic testing at the operational environment in the FDA-regulated industries, this terminology may not be well understood by many software professionals. However, organizations performing regulated operations must be aware of these differences in terminology as they ask for and provide information regarding computer systems.

    •• the new software/hardware or changes can directly affect the management of electronic records

    •• existing hardware/software was never qualified
In some environments, PQ is required, if after performing the OQ in the testing environment, testing software is performed again in preproduction environment

- Conduct an FAT

- Conduct an SAT

- Conduct an IQ

- Conduct an OQ
  ➢ Under certain conditions, an FAT may replace an IQ and/or OQ. Refer to Appendix E.

- Conduct a PQ

- Develop qualification report(s) and/or a validation report
  ➢ Test results and data are formally evaluated

- Conduct a Process/Product PQ. Note: typically, this activity is performed on manufacturing systems only. The objective is to demonstrate that the process consistently produces a result or product meeting predetermined specifications
  ➢ Prepare and approve a Product/Process Summary Report
  ➢ Update any documentation that has references to the application software and its associated version
  ➢ The approved summary report and associated validation documentation are submitted for retention to the National Quality Assurance Document Control Center or site documentation retention center

- *3rd, 4th, and 5th Periods — Early Operational Life, Maturity, and Aging*
  - hardware and software operation and maintenance
  - ongoing reviews/verification program

- *6th Period — Retirement*
  - elimination or replacement of computer systems

SYSTEM DEVELOPMENT FILES

One key element necessary to support the SLC is the availability and maintenance of the system development files. The developer must document the development of each system unit, system component, and configuration item in the software development files and/or use a suitable configuration management tool.

The developer should establish a separate system development file for each unit or logically related group of units; document and implement procedures for establishing and maintaining system development files; and maintain system development files until the system is retired. The system development files should be available for the agency to review upon request; and be generated, maintained, and controlled by automated means. To reduce duplication, each system development file should avoid containing information provided in other documents, or in other system development files. The set of system development files should include (directly or by reference) the following information:

- design considerations and constraints
- design documentation and data
- schedule and status information
- test requirements and responsibilities
- verification and test procedures, and results

Chapter 8

Validation Project Plans and Schedules

REGULATORY GUIDANCE

The validation plan is a strategic document that should state what is to be done, the scope of approach, the schedule of validation activities, and tasks to be performed. The plan should also state who is responsible for performing each validation activity. The plan should be reviewed and approved by designated management.[1]

VALIDATION PROJECT PLANS

Validation plans are documents that tailor a company's overall philosophies, intentions, and strategy to establish performance and computer systems or software adequacy. Validation plans state who is responsible for performing development and validation activities, who identifies which systems are subject to validation, who defines the nature and extent of inspection and testing expected for each  system, and who outlines the framework to be followed to accomplish the validation.

In general, validation project plans describe the organization, activities, and tasks involved in the development of a computer system, including:

- organizational structure of the computerization project
- the departments and/or individuals responsible
- resource availability
- risk management
- time restrictions
- the SLC and development methodology to be followed
- deliverable items
- overall acceptance criteria
- development schedule and timeline
- system release sign-off process
- sample formats for key documentation

The plan may also refer to procedures to guide operation and maintenance.

Validation plans ensure that mechanisms are in place to guide and control the multiple development or maintenance activities, many of which run in parallel, and assure proper

---

[1] FDA, draft of *Guidance for the Industry: 21 CFR Part 11; Electronic Records; Electronic Signatures: Validation,* August 2001.

communications and documentation. During the execution of a project, verification will check the reliability of 'the work' as established by the project plan, so that when the due dates for completion/hand-over arrive, a high degree of certainty exists that the system is validated.

When multiple departments are involved in a project, the system owner will take responsibility for the validation documentation. Other departments will provide documentation and personnel to support the development, validation, and maintenance effort.

Validation plans are not required by any of the predicated regulations but are considered a key project management practice. Validation plans are an essential document for the overall management of projects, and are crucial for the success of the projects.

Typically, managers, their peers, end-users, and those responsible for delivering the system, approve validation plans. Quality assurance may also sign the document. The validation project plan and the requirements specification deliverable, together define the technical and regulatory requirements applicable for a project.

Project validation plans should be started during the early stages of a project. Initial project concepts and planning estimates should be elements in the creation of a project validation plans. The initial project verification activities will assess the project team's capability to produce a validated system and provide input for defining the level of testing effort expected. Project verification will identify any critical deviations to the expected project timing and quality levels, as well as other issues affecting the timely approval of the validation report.

An approved version of the validation plan should be available when a computer technology supplier or contract developer is being selected, and should be updated whenever project events or verification results require a change. An example of a project event would be a change in project scope.

Documents that support the update of the validation plans are:

- system requirements
- criticality and complexity analysis
- project verification results
- other system descriptions

The format of validation project plans is flexible and may incorporate Gantt charts. The contents of the validation plan may include, but are not limited to, the following list:

- document control section
  - system/installation name
  - author(s)
  - creation, save and print date
  - version number
  - document identification
  - document history
    - reviewer and the review date
    - external document references
    - table of contents
  - intended audience
  - scope
  - objective
  - system description
  - validation acceptance criteria
  - verification activities and deliverables

- qualification activities and deliverables
  - ➤ qualification planning
  - ➤ project verification
  - ➤ installation qualification
  - ➤ operational qualification
  - ➤ performance qualification
  - ➤ process validation
- roles and responsibilities
  - ➤ application owner
  - ➤ project management
  - ➤ QA
  - ➤ validation team
  - ➤ computer technology supplier or contract developer
- document control
- project schedule
  - ➤ project activities
  - ➤ project documentation and its delivery

MANDATORY SIGNATURES

Validation plans must be reviewed and approved by the author, the organization owning the execution of the plan, and by the system owner, because they will take the responsibility for the plan.

Validation Plan and Protocols: One document? Many documents?

When developing the validation plan, a decision has to be made about the manner in which the documents will be organised:

- *For small systems, it is possible to integrate the validation plan and all protocols into one document, and to have one report to summarising all results.*
- *For large systems with many components, a validation project plan can be created that divides the validation effort into smaller, more manageable units with separate plans.*
- *Multiple versions of the plan and protocols could be needed, with interim or partial reports.*

The guiding factor for organising validation documentation is the criticality and complexity of the system, and the type of software. Refer to Appendix D.

The review and approval of the plan by QA is optional, but will provide an endorsement that the plan conforms to the current written procedures on computer systems validation, and that the document incorporates applicable regulatory requirements.

PROJECT SCHEDULE

1. Validation plans should include a high-level project schedule that includes the identification of resources. As further project information is known, additional elements can be

incorporated to the validation and project schedule process. Figure 8–1[2] depicts a typical systems development distribution showing the percentage of effort needed for each key developmental activity.

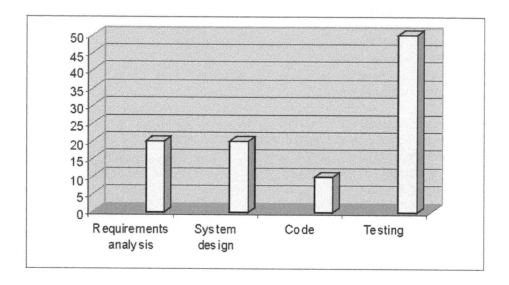

Figure 8–1. Systems Development Distribution.

The following list includes examples of items that need to be tracked as part of the project schedule:

- project kickoff
- validation plan preparation, review, and approval
- training of the validation team
- applicable SLC periods and events
- review and/or development policies, guidelines, standards, and procedural controls
- quality checkpoints for reviews
- project documentation due dates/reviews/approvals
- qualify computer technologies suppliers and/or contract developers (if applicable)
- development and approval of procedural controls
- write specific plans and protocols for IQ, OQ, PQ
- develop the traceability analysis and verify the traceability of the design and testing elements against user requirements
- review/approve specific protocols
- monitor the execution of specific protocols
- assemble, review, and approve the results
- assemble, review, and approve the final validation report

---

[2] Roger S. Pressman, *Software Engineering: A Practitioner's Approach*, 4th ed., McGraw-Hill, NY.

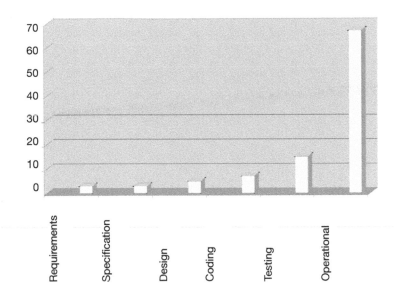

Figure 8–2. Software Development Stages Relative Costs.

Validation plans and associated schedules are live documents that should be reviewed periodically. Phase gate verification activities, performed during each event, may be a perfect place to review the project plan and schedule.

As part of the conceptualization period, senior managers are presented with, and estimate the total effort to be expended (including maintenance for the project). The relative costs of each development stage are depicted in Figure 8–2.[2] *Note*: Maintenance of existing software can account for over 70% of all effort expended by a software organization and this cost will be passed on to the customer.

As part of the schedule, it should be considered that the majority of the development of computer systems fails due to poor gathering of requirements. This situation leads to negatively affect subsequent development activities and associated work products. An investment up front can yield notable savings (cost avoidance) later.

Chapter 9

Inspections and Testing

REGULATORY GUIDANCE[1]

Structural Testing

Structural testing takes into account the internal mechanism (structure) of a system or component. It is sometimes referred to as 'white box' testing. Structural testing should show that the software creator followed contemporary quality standards (e.g., consensus standards from national and international standards development organizations. This testing usually includes inspection (or walk-through) of the program code and development documents.

Static Verifications

These include static analyses such as document and code inspections, walk-through, and technical reviews.

Functional Testing

This testing involves running the program under known conditions with defined inputs, and documented outcomes that can be compared to predefined expectations. Functional testing is sometimes called 'black box' testing.

Program Build Testing

Program build testing is performed on units of code (modules), integrated units of code, and the program as a whole.

INTRODUCTION

Software engineering practices include documented unit testing; code reviews; explicit high-level and low-level design documents; explicit requirements and functional specifications; structure charts and data flow diagrams; function-point analysis; defect and resolution tracking; configuration management; and a documented software development process.

---

[1] FDA, draft of *Guidance for the Industry: 21 CFR Part 11; Electronic Records; Electronic Signatures: Validation*, August 2001.

The FDA expects the same quality principles to be applied during the development and maintenance of computer systems; quality principles must be contained in procedural controls.

Numerous inspection steps undertaken throughout the system development and operational life support that a computer system is validated. These include static analyses such as document and code inspections, walk-through, and technical reviews. These activities and their outcomes help to reduce the amount of system level functional testing needed at the operational environment to confirm that software meets requirements and intended uses.

DOCUMENT INSPECTIONS AND TECHNICAL REVIEWS

Inspections are testing practices in which source code and documentation are examined in either a formal or informal manner in order to discover errors, violations of standards, inconsistencies, and other problems. The objectives of these reviews are to verify the following:

- consistency between the requirements and the specification at each level
- numerical factors
- safety and reliability
- correctness of design
- correctness of sequence
- correctness of formulae and calculations
- traceability to previous design and development levels (i.e., specifications)
- testability
- maintainability
- usability of the product/documentation
- that programming standards, including code layout rules, commenting and naming conventions are consistently applied
- that the code has been produced in accordance with the technical design specification deliverable
- During the development process a set of inspections (i.e., in-process audits) are performed in order to verify the consistency of the code against the design work products. These in-process audits can be used to determine the following:
  - the efficiency of the design algorithms and code produced may be evaluated
  - the consistency of hardware and software interfaces
  - a review of the system requirements and design specification coverage for static and dynamic testing (the software validation plan is the place to specify testing activities)
  - consistency between the system requirements and the technical design
  - consistency between the technical design and the system specification
  - consistency between the code and the technical design description

After technical design has been translated into code,[2] the following inspections should be considered to further evaluate the correctness of the code. A *code review* is an informal white box test where the code is presented in a meeting to project personnel, managers, users, customers, or other interested parties for comment or informal approval. Code reviews are a static test approach that can be applied to source code during the early stage of development.

---

[2] Code or source code (refer to Chapter 16) must be viewed, in this context, in a wide meaning, since it includes, for example, the program listings for custom-built applications; the configurable elements or scripts for configurable applications; or the critical algorithms, parameters, and macros listings for off-the-shelf applications.

Code reviews are performed during software coding and are also applicable during the Operational Life of the application software. They establish that the software can provide all of the features specified during the technical design, with all of the necessary quality attributes, before starting the dynamic testing (unit, integration, and/or black box). It includes:

- verifying the absence of dead code (CPG 7132a.15)
- confirmation of a modular design
- conformance with the required programming approach
- verifying any design mitigation which have resulted from the risk analysis
- verifying that the elements contained in the code will efficiently enable code to be maintained during the its Operational Life

Code inspections are formal testing techniques that focus on error detection. The programmer reads the source code, statement by statement, to a group who ask questions that analyze the program logic, analyze the code against a checklist of historically common programming errors, and verify that the code complies with the appropriate coding standards.

Code audits are formal and independent reviews of source code by a person, a team, or tool, to verify its compliance with software design documentation and programming standards. In addition, the software code is evaluated against its supporting documentation to ensure that it correctly describes the code.

The documentation to be reviewed includes:

- technical design-related deliverable
- system requirements specification deliverable
- system specification deliverables
- test plans
- integration test results
- integration test reports

WHITE BOX TESTING

White box testing (also referred to as structural testing) takes the internal mechanism (structure) of the software into account, and shows that the software creator followed the appropriate quality standards. It usually includes inspection (or walk-throughs) of the program code (code reviews) and development documents.

Code Reviews, Inspections, and Audits

Refer to the previous section.

Unit Testing

The unit test is a white box-oriented test conducted on modules that require a structural test, and is the lowest level of dynamic testing, the testing of an individual program. Individual programs can also be called modules or objects. Unit testing involves the standalone testing of a software unit before it is integrated into the complete software system or subsystem. Any errors found during testing are removed; the test case that uncovered the error (and any associated test cases) is then re-executed. Unit testing may be performed under simulated or emulated conditions. Items that may be checked during the unit testing include the following:

- verification of the functionality as specified in the design document
- verifying the results of subjecting the unit to maximum/minimum limits
- the reaction to incorrect inputs
- exception handling
- algorithm checking
- adherence to documented standards

Code reviews and unit testing are both essential stages of the SLC. If an organization follows established and compliant software quality standards, the result will be more robust programs going into integration testing and, finally, a system with low maintenance. Proper coding and unit testing are basic steps to ensure that the system being built will work, once it is put together.

Less extensive functional testing might be warranted than when the code reviews, inspections, audits, and unit testing are not performed or not available for inspection.

BLACK BOX TESTING

Black box testing is also known as functional testing, benchmark testing and, in the pharmaceutical community, as an operational qualification (OQ). It tests the functionality and correctness of a computer system by running the integrated software, and is performed for one of two reasons: defect detection, and/or reliability estimation. It should cover all functions of the application software that the end-user will use, and will verify the completeness and accuracy of system requirements used to define the test cases.

A black box test involves providing an input to the system and then examining the output. The structural elements of the application software are not examined during the execution of a black box test; these elements are examined during a white box test. An OQ is a black box (result oriented) system test, designed to determine the degree of accuracy with which the system responds to a stimulus. The following sections describe some test cases that an OQ may contain.

Transaction Flow

Transaction flow testing focuses on how transactions are routed within a system, and how queues affect the system's performance. Transactions are measured on the accurate entry, storage, and retrieval of data for various user-controlled operations. Some typical transactions that can be tested include: acknowledgments; receipts and negative acknowledgments; transactions for operational diagnostics; initialization or reset transactions for all external interfaces; transactions utilized during system recovery; transactions used to measure system performance; transactions used to test system security; and recovery transactions from an external system.

It is necessary to identify how the transaction enters the system, how it leaves the system, how it merges (and with whom), how it is absorbed, how it splits, and so on. Queues, as with any resource in a computer system, have a maximum capacity. Exceeding the capacity of a queue can crash a computer system. In addition to the capacity, the testing of a queue includes the order in which transactions are taken off a queue for processing.

Another aspect of transaction flow testing is synchronization. This type of test covers the merging of two or more transactions.

Domain Testing

Domain testing (also known as boundary testing), is used to verify numerical processing software. Domain test values are set at the edges of the valid input ranges (i.e., slightly above and slightly below), and include the biggest, smallest, soonest, shortest, fastest members of a class of test cases. It should be noted that the presence of incorrect inequalities usually causes failures only at the code of algorithm boundaries. The objective of domain testing is to avoid making the computer system 'crash' or involuntarily stop.

State Transitions

Interactive programs move from one visible state to another. If something changes the range of available choices or alters the program display on the screen, the program's state has been changed. A menu system is an example. When an option is selected, the program changes state, and displays a new menu. Each option and pathway in each menu must be tested. State transitions are intended to identify any inconsistencies between the observed functionality and the functionality presented in the system user's guide.

Function Equivalence

This test evaluates the functions embedded in the program, usually performed during the unit testing phase, but for critical functions it can be executed again during the operational qualification. The function in the program being tested is called the test function. The calculation to be compared with the test results can be either calculated by hand or by executing a validated test program.

Load Testing

Load testing is also known as 'stress testing' and includes test cases that cover boundary loading conditions. Stress tests are designed to challenge programs with exceptional situations. This includes not only the volume of data a system may process, but also the rate at which it is processed. A combination of limits is also recommended.

Stress testing executes a system in a manner that demands resources in abnormal quantities, frequencies, or volumes. Some test cases that can be included are: exceeding the average rate of interrupts; increasing the rate of data input; requiring the maximum amount of memory; performing trashing operations in a virtual memory management scheme; performing excessive seeking operations for disk-resident data; and utilizing the maximum number of resources.

Configuration Testing

Configuration testing verifies that the software runs on the specified hardware, using the specified system software and in its specified configuration. Checks for error or warning messages should be carried out.

System Security Testing

This test focuses on policies, procedures, and restrictions relating to both physical access to the system hardware, and to on-line program access.

Regression Testing

The testing of modifications to application software and its associated interfaces must be subject to the same level of testing as performed during software development. The objective is to not only prove the correctness of the modification, but also to uncover any unforeseen effects in other areas of the computer system.

During regression testing, the previous set of tests that were performed successfully are executed again. After correcting an error and installing the revised software, the test cases that caused the error are executed again or the conditions are simulated in order to try and re-create the circumstances that uncovered the error. In some cases a 'fix' will have an effect on something else. It is recommended that time is taken to understand the relationship between the 'fix' and other areas that have already been tested. If necessary, execute the test cases that are related to the affected item again.

Storage Use Testing

For processes in which a high volume of critical data are gathered, it is necessary to determine the adequacy and reliability of the storage system. The adequacy of the storage system is based on the volume of data to be stored and the user requirements.

The FDA publication CPG 7132a.07 provides guidance on the reliability testing associated with computer storage. This guideline requires that computer I/Os are checked for accuracy. This testing is performed in two phases. The first phase consists of qualifying the inputs to the storage system. The second phase requires the implementation of an ongoing monitoring program to periodically verify the I/Os.

Storage use testing is also associated with the Data Archiving Standard Operating Procedure. Primary storage is expensive, which is why data is normally transferred to a less expensive secondary storage media at various times during the execution of the application software. Data transfer procedures should be tested on the guidance provided in FDA publication CPG 7132a.07.

OTHER TESTING TYPES

Program Build Testing

Program build testing, which is also known as integration testing, is the incremental testing performed on units of code (modules), integrated units of code, and the program as a whole in order to uncover any errors associated with either program or hardware interfaces. Unit tested modules are built into a program structure that has been developed to meet the design requirements. The characteristics that are evaluated during the program build testing are:

- interface integrity
- functional validity

- information content

The software interfaces can be internal and/or external to the application program. Internal interfaces are those occurring between the application modules or the hardware components. External interfaces are interfaces to other systems. Guidance on software integration testing is provided in FDA publication CPG 7132a.07. According to this guideline, the data input and output checks for automated systems are necessary to ensure the accuracy on the software interfaces.

Chapter 10

Qualifications

INTRODUCTION

Qualification is the process of demonstrating whether a computer system and its associated controlled process/operation, procedural controls, and documentation are capable of fulfilling the specified requirements. Sidebar 10–1[1] depicts the context of the qualification activities and the computer systems validation process. The physical components of the systems are reviewed and its specific elements, sub-units, and parameters are documented. A qualification also serves to verify and document the acceptability of the design, implementation, integration, and installation of the computer system components. The execution of the qualification is the mechanism by which documented evidence is created that verifies that the computer system performs according to its predetermined specification.

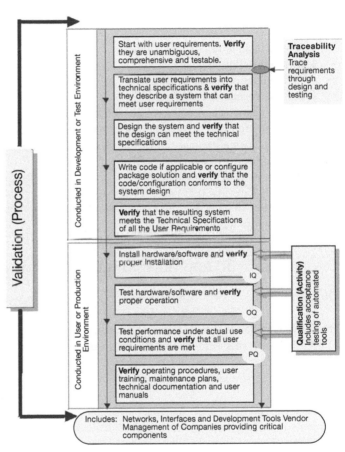

Sidebar 10–1.

Qualification practices apply to all components of a computer system and its associated controlled process/operation, procedural controls, and documentation. Any of these components can be purchased or developed by the company or by a computer technology supplier/contract developer.

The extent of computer technology qualifications depends on a number of factors, including the criticality and the complexity of the system, whether the qualification is to be prospective

---

[1] Courtesy of EduQuest, Inc.

or retrospective, and whether innovative elements are incorporated. Any qualification activity must be conducted before the technology is made available in the operating environment or, in case of manufacturing systems, available for process/ product validation. The qualification must be conducted in operating environment to ensure that the production component, data, and system is not corrupted or compromised.

The qualification of computer hardware, software and the verification of associated documentation are performed according to the applicable sections in Chapter 7.

This chapter describes the general practices for the qualification of computer systems (system software, application software and hardware) that perform regulated operations. The following areas are discussed:

- hardware installation qualification
- software installation qualification
- system operational qualification
- system performance qualification
- operating systems and other system software qualification
- standard instruments, microcontrollers, and smart instrumentation qualification
- standard software packages or off the shelf software qualification
- configurable software qualification
- custom-built systems qualification

The information contained in this chapter is written by enumerating the key practices that relate to specific qualifications.

HARDWARE INSTALLATION QUALIFICATION[2]

Objective

Hardware installation qualification of the computer technology provides documented evidence of a high degree of assurance that the hardware has been installed according to an appropriate and approved technical design specification, and the hardware supplier's installation manual(s). Appropriate personnel must conduct the installation qualification of the computer systems hardware.

The term 'computerized hardware' (hereafter referred to as hardware) refers to physical elements including input devices, output devices, network components, signal converters and computers (e.g., controllers, microprocessors, smart instruments).

Practices

The following practices are applicable to all hardware components considered part of a computer technology performing regulated operations:

Installation of hardware components that are part of a computer system are qualified according to procedural controls.

Requirements for verifying and recording the results of the installation of computer hardware must be contained in procedural controls.

---

[2] O. López, *Regulatory Systems Hardware Qualification*, Sue Horwood Publishing Limited, ISBN 1–904282–05–9 (www.suehorwoodpubltd.com).

Hardware installation test cases may be developed with the collaboration of the application support personnel, the application expert, or system user.

Equipment separate from the computer system but considered part of the computer system must be determined (if calibration and qualification is required) prior to, or as part of, the computer system installation qualification; this situation may arise in certain laboratory or shop floor situations.

If calibration of ancillary equipment or tools used for installation is required, then the measurement methods, traceability of the calibration standard to international, national, or approved standards, and equipment specifications must be provided for acceptance.

If a validation plan is available, the hardware installation qualification should be addressed in this plan. If there is no validation plan, the hardware installation plan may be addressed either in a standalone document or as part of the installation qualification protocol. The hardware installation qualification plan identifies the activities to be completed and the responsible parties involved.

Hardware installation activities and documentation must be clearly communicated to the hardware/software supplier(s) before reaching any contractual agreements.

Before the execution of the hardware IQ, checklists, tables, and/or forms must be created to collect the test result data. These are approved as part of the IQ and completed during the qualification. The IQ protocol covers all installation activities and may include, but not be limited to, the following:

- bill of materials
- environmental measurements
- design specifications and acceptance criterion
- verification of the as installed diagrams
- installation verifications
- date and time of execution
- personnel performing the qualification/verifications
- signatures of installation personnel
- acceptances signatures of appropriate personnel

The inspection of all hardware must be indicated and documented, including the equipment received, the personnel performing the activity, and the date of the activity. A complete list of the hardware to be qualified must be created, including the hardware item, description, manufacturer or supplier, manufacturer's model number, and serial number.

Hardware configuration diagrams must be retained as part of the qualification documentation. This is especially important in certain laboratory and shop floor situations where the computer hardware is essential to the operation of the equipment.

A checkout of critical input/output and electrical connections must be documented, including (as applicable):

- wiring connections and I/Os
- radio frequency transmission devices
- shielding and wire type specifications
- grounding
- critical signal levels
- verification of 'as-built' wiring diagrams

Electrical specifications must be documented and verified, including:

- voltage
- amperage
- phase requirements
- power conditioning requirements
- backup power-supply requirements

For computer workstations, the workstation configuration must be verified using the technical design specification. The result of the verification must be documented using the data collection forms located in the approved protocol.

Environmental requirements and specifications must be documented and assurance provided that the environmental parameters are within the range specified by the manufacturer.

Critical environmental requirements may exist in the following areas:

- temperature (e.g., air, fluid, hydraulic, etc.)
- humidity
- electrical enclosure ratings
- airflow
- particulate control
- pressure
- cleaning requirements
- spurious radio frequency interference
- electromagnetic interference

Verification must be provided for the approach utilized, to mitigate any items identified in the risk analysis that is related to the hardware.

The physical security arrangements for the hardware must be verified. Critical hardware must be placed in a physically secure area to prevent any unauthorized physical access.

It must be verified that the hardware installation adheres to the applicable safety standards and specifications in the areas of:

- fire safety
- OSHA regulations
- electrocution hazards
- ergonomic requirements

Verify that the equipment has been tagged or identified according to the applicable policies and procedures.

Verify spare parts lists (if applicable).

Verify the hardware specifications for routine and preventive maintenance.

Execute any supplied diagnostics to ensure proper installation.

The installation qualification will include, but will not be limited to, the following technologies/documentation:

- infrastructure
- storage devices
- colour graphics and displays
- printers
- hard disk and floppy systems
- I/O termination units
- signal conditioners and transducers

- sensors and control elements
- field devices
- electrical services (e.g., power sources, UPS, grounding, shielding)
- as built drawings
- supplier SLC documentation supplied under the contract
- customer documentation (e.g., user requirements, validation plan, supplier audit, and supplier selection, as applicable)

Once the hardware installation qualification protocol has been completed, the test results, data and documentation are formally evaluated. The written evaluation should be presented clearly and in a manner that can be readily understood. The structure of the report can parallel the structure of the associated protocol. The report should also address any nonconformances encountered during the hardware installation qualification, and their resolution. The hardware installation qualification report summarizes the results of the verification and testing of all hardware technologies that are part of the system.

The Hardware Installation Qualification Report can either be a standalone document or may be incorporated into the overall system qualifications summary report.

The approved Hardware Installation Qualification Report should be submitted for retention to the appropriate record retention organization.

Networks are generally extensions of distributed processing. They may consist of connections between complete computer systems that are geographically distant, or they may consist of computer systems on a LAN in the same facility.

In *Guide to Inspection of Computerized Systems in Drug Processing*, the FDA raises the following issues concerning networks:

- What outputs, such as batch production records, are sent to other parts of the network?
- What kinds of input (e.g., instructions, software programs) are received?
- What are the identities and locations of establishments that interact with the firm?
- What is the extent and nature of the monitoring and controlling activities exercised by remote on-net establishments?
- What security measures are used to prevent unauthorized entry into the network, and what are the possibilities for unwarranted process alterations, or the obliteration of process controls and records

For network implementation purposes, the first decision is whether the network is considered an open or a closed system. The open/closed system issue will drive the technological and procedural controls that need to be considered during the system implementation, and that need to be addressed as part of the system risk analysis.

SOFTWARE INSTALLATION QUALIFICATION

Objective

The installation qualification performed on the applications and the system software provides documented evidence to show, with a high degree of assurance, that the software has been installed according to an appropriate and approved technical design specification, and to the supplier's installation manual(s). Appropriate personnel must conduct the installation qualification of the application and system software.

Table 10–1. Software IQ Deliverables and Acceptance Criteria.

Deliverable	Acceptance Criteria
Criticality analysis report	verified and approved document
System requirements	verified and approved as-built document
System specification	verified and approved as-built document
Technical design specification	verified and approved as-built document
Traceability analysis	verified and approved as-built document that traces system requirements to specific section/clauses in the system specification, technical design specification, development testing, and acceptance testing documentation
Assessment(s) and progress audit reports	verification that the SLC documentation conforms to the user's requirements and the supplier's quality procedures
Test plan	verified to be in conformance with the as built system
Training plan	verified and approved for execution
System environment and related utilities test results	installed as designed
System physical security	installed as designed
User guides and manuals	verified and approved, as-built
Technical guide	verified and approved, as-built
Installation and maintenance instructions	verified and approved, as-built
Completed hardware installation	installed as designed, and successful startup and shutdown of the computer system
System software installation	installed as designed, and successful startup and shutdown of the software
Program source code (if applicable)	conform copy of as delivered, and successful startup and shutdown of the software
System configuration	installed as designed
Application installation and inventory	installed as designed
Initial system data (data conversion, if applicable)	installed as designed
System management procedural controls	verified and approved as-built
Maintenance and log book	installed as designed
Service level agreements	verified against the contract and approved

Practices

The following practices are applicable to all software components considered part of all computer systems performing regulated operations:

The installation of software components that are part of a computer system are qualified according to procedural controls.

The requirements for verifying and recording the results of the installation of computer application and system software must be contained in procedural controls.

The software installation test cases may be developed with the collaboration of the application support personnel, the application expert, or a system user.

If a validation plan is available, the software installation qualification should be addressed in this plan. If there is no validation plan, the software installation plan may be addressed in either a standalone document or as part of the installation qualification protocol. The software installation qualification plan identifies the activities to be completed and the responsible parties.

Table 10–1 lists the deliverables to be gathered and verified during the installation of system and application software. The results of this activity must be recorded in the IQ protocol or similar document.

The list below also includes some brief acceptance criteria, but this should not necessarily be considered to be a complete description.

The software installation qualification can be executed in parallel with, or after, the hardware installation qualification.

If applicable, software 'virus checking' must occur before the installation qualification of the software.

Before the execution of the software IQ, checklists, tables, and/or forms must be created to collect the test results data. These are approved as part of the IQ protocol and completed during the qualification. The IQ protocol covers all of the installation activities and may include, but should not be limited to, the following:

- personnel performing the activities
- computer hardware and software installation activities
- recording of software and hardware configuration parameters
- verification of installation requirements, including database structure, files, tables, and dictionaries, as applicable
- provision of appropriate utilities to demonstrate the correct installation/connectivity
- verification of environmental and power utilities
- complete list of the software to be qualified is created including:
  - name and version of the product
  - brief description of the software (e.g., operating system, document management application package, custom application, etc.)
  - suppliers' names
  - license numbers or other identifying information (if applicable)

Prior to software installation, a 'back-out' or 'roll-back' plan and procedure must be documented. If the planned installation fails, the 'back-out' plan is implemented and the restoration activities and results are documented. The 'back-out' plan includes information on performing a 'backup' of the system, and/or making use of an existing 'backup' of the system, which has been verified prior to software installation.

Upon successful installation of the software, a printout of the resulting directory listings may be obtained for verification and documentation purposes. This printout may be enclosed as part of the executed software IQ protocol.

If possible, a printout of the installation 'audit' or 'system log' file should be captured at the time of software installation and retained as documented evidence. In some cases, the supplier may provide internal diagnostic software that verifies the successful installation/connection of the computer technologies. A printout of any such diagnostic test results or reports should be retained as evidence. If it is not possible to capture this evidence, then the expected results of a successful installation must be indicated in the IQ protocol, and the actual observed results documented.

All installation qualification activities and resulting data must be verified by a second (independent) person. Evidence of verification activities may be confirmed by signing of the corresponding data collection form; verification activities must include the date on which the activity was performed. The reviewer's signature indicates that the test was completed as expected, that the acceptance criteria were met, and that the appropriate documentation/ evidence (if applicable) was collected.

If nonconformances or deviations to the project plan or IQ protocol are encountered during the software IQ testing, these must be documented, analyzed, resolved, reviewed, and approved. The resolution process must indicate what additional actions must be taken to provide a conforming product (e.g., return the program to development for error analysis and correction, and re-execute the test script after correction; the nonconformance was due to an inaccuracy in the test script, review and update the test script, etc.). After the successful resolution of the non-conformance has been obtained, the original test, the nonconformance information, and the retest must all be retained, approved, and reviewed by the appropriate personal.

Once the software installation qualification protocol has been completed, the test results, data, and documentation are formally evaluated. The written evaluation should be presented clearly and in a manner that can be readily understood. The structure of the report can parallel the structure of the associated protocol. The report should also address any nonconformances encountered during the software installation qualification and the resolution. The software installation qualification report summarizes the results of the verification and testing of all technologies that are part of the system.

The Software Installation Qualification Report can either be a standalone document or may be incorporated into the system qualification summary report.

The approved Software Installation Qualification Report should be submitted for retention by the appropriate records retention organization.

SYSTEM OPERATIONAL QUALIFICATION

Objective

The system operational qualification performed on the computing technology provides documented evidence of a high degree of assurance that the hardware and software components perform as required by the system specification. This includes the verification that each unit or subsystem of the system operates as intended throughout all anticipated operating ranges. Appropriate personnel must conduct the system operational qualification.

Practices

The following practices are applicable to all system components considered part of computer technologies which perform regulated operations:

The system components that are part of the computer technologies are qualified according to procedural controls.

The requirements for testing and recording the results of the operation of the computer technologies must be contained in procedural controls.

The system operational test cases may be developed with the collaboration of the application support personnel, the application expert, or a system user.

If a validation plan is available, the system operational qualification should be addressed in this plan. If there is no validation plan, the system operational qualification plan may be addressed in either a standalone document or as part of the system qualification protocol. The system operational qualification plan identifies the activities to be completed and the responsible parties involved.

Before execution of the operational qualification, an installation qualification must be performed on all relevant computer technologies.

The application support personnel, the application expert, or a system user may develop the system operational qualification test cases.

Components that require calibration are identified and calibrated by qualified personnel before, or as part of, the installation qualification.

Before the execution of the system OQ, checklists, tables, and/or forms must be created to collect the test results data. These are approved as part of the OQ protocol and completed during the qualification. The OQ protocol covers all of the system operational activities and may include, but should not be limited to, the following:

- personnel performing the activities
- testing of the software and hardware configurable parameters
- the use of any data migration tools. These should be written in accordance with an established quality management system. The data migration process should be verified and the verification performed before final data loading for release
- verification of applicable functions:
- verification of the executed computer technologies installation qualification
- audit trials
- electronic signatures
- alarms, alarms actions, events, interlocks, and safety circuit testing
- confirmation of data acquisition
- algorithm, data handling, and report generation
- timers (if applicable)
- power failure and recovery
- disaster recovery
- proper communications between the computer system and its peripherals
- restart and recovery functions and procedures
- availability of an application for authorized accounts holders
- integration with other computer system

As explained in Appendix E, the content of the FAT, SAT, and OQ can be similar. The system specification forms the basis for these activities.

Taking into consideration the previous statement, for the purpose of project management the operational qualification may be divided into two (2) main types: the OQ1 or FAT, and OQ2 or SAT. See Figure 10–1. Additional information can be found in Appendix E.

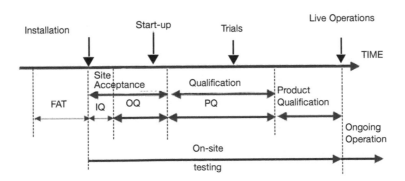

Figure 10–1. Qualification Timeline.

In very large and complex systems developed by contract developers, it is the practice to perform an FAT or OQ1. The FAT is performed in the development environment by the supplier/integrator on the supplier's/ integrator's premises in order to verify the proper operation of the application before delivery to the user's site. If the FAT results comply with the system specification, the system is released for shipment and installation on the user's site.

All critical deviations found during the execution of the FAT must be corrected before system release and shipment.

Regression testing must be executed in order to verify that all critical deviations were corrected.

In all large and complex systems developed by contract developers, it is normally the practice to perform an SAT. The SAT is the set of activities comprising the system installation, start-up, operational testing, and user orientation performed by the software supplier/integrator in the operational environment. For the purposes of this book, the operational testing performed as part of the SAT is also known as the OQ2. If the SAT is conducted according to approved protocols, test results are properly documented, and test results reviewed and approved by the user and QA, it may not be necessary to repeat the installation and operational qualifications. This provides cost savings and can streamline the validation process. Refer to Sidebar 10–2.

All system operational qualification activities and resulting data must be verified by a second (independent) person. Evidence of verification activities is confirmed by signing of the corresponding data collection forms. The verification activities must include the date that the activity was performed; the reviewer's signature indicates that the test was completed as expected, that the acceptance criteria were met, and that the appropriate documentation/evidence (if applicable) was collected.

If nonconformances or deviations to the plan or protocol are encountered during the system OQ testing, these must be documented, analyzed, resolved, reviewed, and approved. The resolution

---

THE OQ (1) and (2) CONCEPT

OQ_1 refers to the testing performed in a development environment, which can be on the suppliers' premise, or within the user's company. The supplier/integrator performs this testing and it is witnessed by the company personnel, including QA. OQ_1 is often referred to as an FAT.

OQ_2 refers to the testing performed in an operational environment, usually the environment where the system will be used when operational, but before official release. The supplier/integrator performs this testing and it is witnessed by the company personnel, including QA. OQ_2 is often referred to as a SAT.

This concept applies to all technologies of a computer system.

APPLYING THE OQ's (1) and (2) CONCEPT

Since OQ_2 witnessed testing takes place under operational conditions, there is no need to give guidance on the control of the testing environment. The IQ will provide evidence that the installation is appropriate for operational use.

During the execution of OQ_1, in the development environment, it should be ensured that this environment models the specified operating environment. This is because unrealistic or unconvincing test environments can make tests unsuitable as evidence of proper operation. Unrealistic tests could be tests that are performed on old program versions, or where there is no evidence of regression testing for the program. When consolidating of OQ_1 and OQ_2 tests, only the tests that have been identified as suitable should be taken into account.

Sidebar 10–2.

---

process must indicate what additional actions must be taken to provide a conforming system (e.g., return the program to development for error analysis and correction, and re-execute after correction; the nonconformance was due to an inaccuracy in the test script; update the test script; and so on). After the successful resolution of the nonconformance, the original test performed, the nonconformance information obtained, and the retest must all be retained, approved, and reviewed by the appropriate person.

Once the system operational qualification protocol has been completed, the test results, data, and documentation are formally evaluated. The written evaluation should be presented clearly and in a manner that can be readily understood. The structure of the report can parallel the structure of the associated protocol. The report should also address any nonconformances encountered during the system operational qualification and their resolution. The system operational qualification report summarizes the results of the verification and testing of all technologies that are part of the system.

The System Operational Qualification Report can either be a standalone document or may be incorporated into the overall system qualification summary report.

The approved System Operational Qualification Report should be submitted for retention to the appropriate records retention organization.

Upon successful completion of the operational qualification, the computer system is available for further qualification (PQ).

SYSTEM PERFORMANCE QUALIFICATION

Objective

Computer systems performance qualification performed on the computing technology provides documented evidence of a high degree of assurance that the computer system and the operating environment (Figure 2–1) perform as established by the requirements specification deliverable.

Appropriate personnel must conduct the system performance qualification.

Practices

The following practices are applicable to the integrated computer system (which include the controlled process, operating procedures, and supporting documentation) that performs regulated operations:

- The integrated computer system is qualified according to procedural controls
- The requirements for testing and recording the results of the integration of the computer system must be contained in procedural controls
- The system integration test cases may be developed with the collaboration of the application support personnel, the application expert, or a system user

If a validation plan is available, the system performance qualification should be addressed in this plan. If there is no validation plan, the system performance qualification plan may be addressed in either a standalone document or as part of the system qualification protocol. The performance qualification plan identifies the activities to be completed and the responsible parties involved.

The application support personnel, the application expert, or a system user may develop the system performance qualification test cases.

Before the execution of the system performance qualification, checklists, tables, and/or forms must be created to collect the test results data. These are approved as part of the PQ protocol and completed during the qualification.

The specific objectives of the system performance qualification include:

- determining the accuracy of the integrated system in the receiving, recording, storing, and processing of data
- determining the accuracy of the integrated system in arriving at an appropriate decision based upon the data received and the established decision matrix

For a process control system, the system performance qualification may involve simulation run(s) on the controlled process and its associated equipment using placebo or process materials. These runs are not the same as the product/process performance qualification, although they can be used to support this activity. System performance qualification demonstrates and verifies computer system suitability. Product/process performance qualification demonstrates and verifies product/process suitability.

For business/corporate computer installations, it may be difficult to separate system operational qualification and system performance qualification. Procedural controls may be used to address this issue.

All system performance qualification activities and resulting data must be verified by a second (independent) person. Evidence of the verification activities are confirmed by signing the corresponding data collection form. The verification activities must include the date that the activity was performed. The reviewer's signature indicates that the test was completed as expected, that the acceptance criteria were met, and the appropriate documentation/evidence (if applicable) was collected.

If nonconformances or deviations to the project plan or protocol are encountered during the system PQ testing, these must be documented, analyzed, resolved, reviewed, and approved. The resolution process must indicate what additional actions must be taken to provide a conforming system (e.g., return the program to development for error analysis and correction, and re-execute the test script after correction; the nonconformance was due to an inaccuracy in the test script, update the test script, etc.).

After the successful resolution of the nonconformance, the original test performed, the non-conformance information obtained, and the retest must all be retained, approved, and reviewed by the appropriate personal.

Once the system performance qualification protocol has been completed, the test results, data, and documentation are formally evaluated. The written evaluation should be presented clearly and in a manner that can be readily understood. The structure of the report can parallel the structure of the associated protocol. The report should also address any nonconformances encountered during the system operational qualification and their resolution. The system performance qualification report summarizes the results of the verification and testing of all technologies that are part of the integrated system.

The System Performance Qualification Summary Report can either be a standalone document or may be incorporated into the overall system qualification summary report. The approved System Performance Qualification Summary Report should be submitted for retention to the appropriate records retention organization.

Upon successful completion of the performance qualification, process control systems are available for process/product performance qualification and business computer systems are available for operation.

OPERATING SYSTEM AND UTILITY SOFTWARE INSTALLATION VERIFICATION

Introduction

The practices contained in this section apply to all operating systems and system software components that are considered part of a computer system.

System software (e.g., operating systems, assemblers, utilities) is defined as:

- independent software that supports the running of application software (ISO)
- software designed to facilitate the operation and maintenance of a computer system and its associated programs (IEEE)

Objectives

All computer systems performing regulated operations must be validated. To support computer systems validation, operating systems and system software components are installed according to the manufacturer's instructions. The installation verification includes documentation gathered during the verification activities.

Practices

The installation of system software and its associated components is verified according to normal software installation qualification practices. (Refer to *Software Installation Qualification* in this chapter.)

Well-known operating systems do not need functional testing to be performed by the end-user.

Due to the close relationship between the computer hardware and the system software, operating system installation verification may be performed during the hardware installation qualification activities.

New versions of system software (e.g., operating systems, assemblers, and utilities) should be reviewed prior to use and consideration should be given to the impact caused by any new, amended, or removed system software features on the application(s) that run on it. This impact assessment could lead to formal regression testing of the application(s), particularly when a major upgrade of the operating system has occurred. The results of the regression testing must be documented and approved by the appropriate personnel:

The following project activities described in Chapter 7 are also applicable to system software and system software components:

- Conduct a system (hardware, software and process) risk study and a criticality/complexity assessment. Usually this study and assessment are performed to quantify the impact of new versions on the actual configuration
- Complete change control forms and approvals, including obtaining the approval of QA
- Preparation of the installation qualification protocol, if applicable, which includes activities associated with the regression testing of the applications impacted by the installation of system software
- Execute the IQ protocol

- If applicable, perform regression testing of the application programs (OQ and PQ)
- Prepare a summary report
- Update the documentation regarding references to system software, system software components, and their associated versions
- The approved summary report and associated validation documentation should be submitted for retention to the appropriate records retention organization

Part 11 Areas of Interest

The system administrator's position requires the ability to access all items in the operating system. Key areas of concern are the protection of records, the integrity of the application software files, and the operating system configurable parameters. In addition to the security of the system, the Part 11 areas to be considered for an operating system are as follows:

- Are the records protected against intentional or accidental modification or deletion?
- Is an electronic audit trail function automatically generated for all system administrator entries?
- Is the audit trail completely outside of the control and access of the system administrator (except for the read-only access of the audit trail file)?
- Is it impossible to disable the audit trail function without being detected?
- Is the system date and time protected from unauthorized change?
- Is the audit trail data and/or metadata protected from accidental and/or intentional modification or deletion?
- Is the distribution of, access to, and use of operating system and maintenance documentation controlled?

STANDARD INSTRUMENTS, MICROCONTROLLERS, SMART INSTRUMENTATION VERIFICATION

Introduction

The practices contained in this section apply to all standard instruments, microcontrollers, and smart instrumentation (e.g., weigh scales, bar code scanners, controllers, vision systems and EPROM's) considered part of a computer system. This equipment is driven by programmable firmware.

Objective

All computer systems performing regulated operations must be validated. To support computer systems validation, standard instruments, microcontrollers, and smart instrumentation are functionally tested (black box test) and the installed versions recorded.

Practices

The installation of standard instruments, microcontrollers, smart instrumentation, and its associated components is verified according to normal software installation qualification practices. (Refer to *Software Installation Qualification* in this chapter.)

Since the user's company does not have control over the system product to be acquired, the user's company must adequately understand and specify the requirements and capabilities to be provided by the system product.

Standard instruments, microncontrollers, and smart instrumentation can be configurable. The configuration must be recorded using an IQ protocol, including the version of the hardware, versions of the firmware (if applicable) and associated software.

The introduction of new versions of firmware or software must be performed through the completion of rigorous change control.

Backups of firmware or software are critical items for operational continuity.

The impact of new versions of firmware and software on the validity of the existing qualification documentation should be reviewed and appropriate action taken. If applicable, after the new version is installed, regression testing should be conducted. The results of the regression test must be documented and approved by the appropriate personnel.

If a manual system is being replaced by one of these technologies, the two systems should be run in parallel for a time as a part of the performance qualification.

The operational qualification for standard instruments, microcontrollers, and smart instrumentation consists of a 'black box' test. This type of test is based on the user's firm application requirement and challenges a program's external influences. It views the software as a black box concerned with program inputs and its corresponding outputs. The 'black box' testing must consider not only the expected (normal) inputs, but also unexpected inputs. Black box testing is discussed in Chapter 9.

If data is recorded onto magnetic media (electronic records), then the following will apply:

- Records should be secured by physical or electronic means against wilful or accidental damage. The records collected should be checked for accessibility, durability, and accuracy. If changes are proposed to the computer equipment or its programs, the above mentioned checks should be performed at a frequency appropriate to the storage medium being used.
- Records should be protected by backing up at regular intervals. The backups should be stored for as long as is necessary to conform to regulatory and company requirements, and at a separate and secure location.

There should be adequate alternative arrangements made available for systems in case of a breakdown. The time required to bring the alternative arrangements into use should be related to their criticality and the possible urgency of the need to use them.

The procedural controls to be followed if the system fails or breaks down should be defined and qualified. Any failures and remedial actions taken must be recorded.

Procedural controls must be established to record and analyse errors, and to enable corrective action to be taken.

The following project activities described in Chapter 7 are also applicable to standard instruments, microcontrollers and smart instrumentation:

- Define the scope of the system and its strategic objectives
- Perform a preliminary evaluation of the technology
- Provide a Process Description including process flow diagrams and process and instrumentation drawings (P&IDs)
- Complete the change control forms and approvals, including obtaining the approval of QA
- Prepare a system requirement specification deliverable
- Prepare and approve a Validation Plan

- Conduct a System (hardware, software and process) risk study and a Criticality/Complexity Study and document the results. Any new requirements that result from the risk study must be added to the system requirements specification deliverable
- Approve the system requirements specification deliverable
- Select the supplier
- Revisit the risk study
- If changes to the hardware and/or software are required, develop and approve the hardware/ software design, implement the hardware and software using the design specification, and conduct an integration test
- Revisit the risk study to confirm that any risks have been addressed
- Prepare validation/qualification protocols
- Conduct an IQ
- Conduct an OQ/PQ (usually these can be carried out together)
- Develop and approve a summary report
- If applicable, and as part of the manufacturing system's interface with the standard instruments, microcontrollers, or smart instrumentation, conduct a Process/Product Performance Qualification. *Note*: This activity is typically performed on manufacturing systems only. Prepare a Product/Process Report
- Update the documentation regarding references to application software and their associated version
- The approved summary report(s) and associated validation documentation should be submitted for retention by the appropriate record retention organization

Part 11 Areas of Interest

Standard Instruments, Microcontrollers, and Smart Instruments are usually connected to a device to record their data. It is uncommon for them to contain functionality that supports electronic signatures. The majority of these systems are hybrid systems and if this is the case, the Part 11 areas to be assessed include:

- audit trails and metadata management
- system security
- access code and password maintenance and security
- password assignments
- operational checks
- authority checks
- location checks
- document controls
- records retention and protection
- open/closed system issues

On hybrid systems, another area to be evaluated is the procedural controls on the link between handwritten, signed (paper-based) report(s) and the electronic records associated with the report. This link to the record(s) must be based on a unique attribute.

The areas of Part 11 that need to be assessed for Standard Instruments, Microcontrollers, and Smart Instruments when these devices are not connected to a recording device include:

- system security

- access code and password maintenance and security
- password assignments
- operational checks
- authority checks
- location checks
- document Controls
- open/closed system issues

STANDARD SOFTWARE PACKAGES QUALIFICATION

Introduction

Standard Software Package are products that have already been developed and are usable as is or with modifications.

The practices contained in this section apply to Standard Software Packages (e.g., spreadsheets, text processors, database programs, CAD systems, graphics packages, and programs for technical or scientific functions) that are considered part of a computer system. By definition, this includes noncustomized application software that has been purchased and that has not been modified.

When classifying a software application as either a Standard Software package or Configurable Software package, consideration should be given the amount of application development needed for the proposed solution. The classification selected will determine the validation approach to be taken. For example, SAS is a comprehensive data management system with modules for application development, spreadsheets, graphing, data entry, data analysis, project control, business process management, and database query. If the SAS programs are too complex, SAS may be considered to be a Configurable Software package instead of a Standard Software package.

Another item of concern regarding Standard Software Packages is that the package by itself may not be Part 11 compliant. An example of this is a package that does not have Part 11-compliant audit trails functionality. In this case, the Standard Software package may need to be combined with third-party applications and/or plug-ins, in order to make it compliant with Part 11.

A reference to source for the testing of standard software packages is ISO/IEC 12119, *Information technology – Software packages – Quality requirements and testing*. It is applicable to standard software packages and it provides the following:

- a set of quality requirements for standard software packages
- instructions on how to test standard software package against specified quality requirements
- a quality system for a supplier who is outside of the scope of ISO/IEC 12119

A RELATED PRODUCT FOR ISO/IEC 12119, THE IEEE STANDARD ADOPTION OF ISO/IEC 12119

Objective

All computer systems performing regulated operations must be validated. To support computer systems validation, the applications developed using standard software packages are functionally tested (black box testing) and the installed versions recorded.

Practices

Standard software packages are called canned or COTS (Commercial Off-The-Shelf) configurable packages. The qualification of standard software packages is performed according to normal software installation and operational qualification practices. (Refer to *Software Installation Qualification* and *System Operational Qualification* in this chapter.)

Since the user's company does not have control over the requirements of the standard software package, the user's company must identify the application requirements and match them to the available standard software packages.

There is no requirement to validate the standard software package. The validation concentrates on the elements developed to support the application.

The linking and/or configuration of the standard software package modules is used to develop the final application.

Standard software packages are configurable, and this configuration must be recorded in the equipment IQ, along with the version.

Change control must be rigorously applied, since changing these applications is often very easy, and they have limited security.

The impact (criticality) of new versions on the validity of the qualification documentation must be reviewed and appropriate action taken. If applicable, after a new version has been installed, the test cases contained in the operational qualification should be executed again.

If a manual system is being replaced, the two systems should be run in parallel for a time, as a part of the operational qualification.

The qualification effort should concentrate on the application itself and should include:

- application requirements and functionality
- the high-level language, scripts, or macros used to build the application
- critical algorithms, formulae and parameters
- data integrity, security, accuracy, and reliability
- operational procedural controls

User training should emphasize the importance of change control and the need to maintain the integrity of these systems.

Data shall be secured by physical and/or electronic means against willful or accidental damage. Stored data must be periodically checked for accessibility, durability, and accuracy. If changes are proposed to the computer equipment or its programs, the checks mentioned above should be performed at a frequency appropriate to the storage medium being used.

Data must be protected by backing up at regular intervals. Back-up data must be stored for as long as is necessary to comply with regulatory and company requirements at a separate and secure location.

There should be adequate alternative arrangements made available for systems in case of a breakdown. The time required to bring the alternative arrangements into use should be related to their criticality and the possible urgency of the need to use them.

The procedural controls to be followed if the system fails or breaks down should be defined and training provided. Any failures and remedial actions taken must be recorded.

Procedural controls must be established to record and analyse errors, and to enable corrective action to be taken.

The operational qualification for standard software packages consists of a 'black box' test. This type of test is based on the user's firm application requirements specification deliverable and challenges the packages external influences.

The following project activities described in Chapter 7 are also applicable to standard software packages:

- Define the scope of the system and its strategic objectives
- Perform a preliminary evaluation of technology
- Complete the change control forms and approvals, including obtaining the approval of QA
- Prepare a system requirement specification deliverable
- Prepare and approve a Validation Plan
- Conduct a System (hardware, software, and process) risk study and a Criticality/Complexity Study and document the results. Any new requirements that result from the risk study must be added to the system requirements specification deliverable
- Approve the system requirements specification deliverable
- Select the software package
- Revisit the risk study
- Prepare and approve the system specification deliverable in full
- Prepare a technical design specification deliverable
  - critical algorithms, parameters, and formulae
  - data integrity, security, accuracy, and reliability
- Perform design reviews, including technical design inspections
- Approve the technical design specification deliverable
- Prepare validation/qualification protocols and the applicable procedural controls
- Build the application
- Inspect the developed code. The focus of this inspection is on the critical algorithms and parameters
- Conduct an IQ
- Conduct an OQ and include in this process the operation of the application using the procedural controls
- Prepare a summary report
- Update the documentation regarding references to application software, associated versions, and user manuals

The approved summary report(s) and associated validation documentation should be submitted for retention to the appropriate record retention organization.

Part 11 Areas of Interest

The output from standard software packages is typically recorded to a magnetic media. Part 11 areas to be assessed include:

- audit trails and metadata management
- system security
- access code and password maintenance and security
- passwords assignments
- operational checks
- authority checks
- location checks
- document controls
- records retention and protection
- open/closed system issues

If electronic signatures are implemented, the following additional areas must be assessed:

- electronic signature security
- electronic signature without biometric/behavioral detection
- electronic signature with biometric/behavioral detection
- signature manifestation
- signature purpose
- signature binding
- certification to FDA

CONFIGURABLE SOFTWARE QUALIFICATION

Introduction

The practices contained in this section apply to Configurable Software (e.g., SCADA systems, Laboratory Information Management Systems, Manufacturing Resource Planning Systems, Document Management Systems, Building Automation/Management Systems, Electronic Batch Recording Systems, Electronic Document Management Systems, Manufacturing Execution Systems, etc.) that are considered part of a computer system. The level complexity of the individual 'programs' that are used to form the released application are a critical factor to consider for this type of software package.

Objective

All computer systems performing regulated operations must be validated. To support computer systems validation, the quality of the software embedded in the standard product must be established. The configurable elements of the standard product must be reviewed and tested.

Practices

Configurable Software packages are also called custom configurable packages.

The software system and hardware platform must be well known and mature before being considered to be Configurable Software; otherwise the software validation activities for Custom-Built systems must be used.

A typical feature of configurable software packages is that they permit users to develop their own applications by configuring/amending/scripting predefined software modules and by developing new application software modules using the standard (core) product. Each application (of the standard product) is therefore specific to the user's process and maintenance becomes a key issue, particularly when new versions of the standard product are released.

Since the user's company does not have control over the requirements of the standard product, the user's company must identify the application requirements and match these to the standard product that are available.

There is no requirement to validate the standard product but the quality embedded in the standard product must be evaluated as follows:

- Conduct a quality audit of the supplier of the standard product to review the level of quality and structural testing built into the standard product
- If the end-user cannot directly review the level of quality embedded in a standard product, extensive functional testing must be performed in order to assess the quality of the standard product

The review activities for the standard product should concentrate on the following:

- Define the scope of the system and its strategic objectives
- Perform a preliminary evaluation of technology
- Conduct a feasibility study
- Generate a process description which includes process flow diagrams/data flow diagrams and, if applicable, PI&Ds
- Complete the change control forms and approvals, including approvals of QA
- Develop system requirements specification deliverable
- Prepare and approve a Validation Plan
- Conduct a system (hardware, software, and process) risk study and a Criticality/Complexity Study and document the results. Any new requirements that result from the risk study must be added to the requirements specification deliverable
- Approve the system requirements specification deliverable
- Conduct a quality audit of the software supplier (main candidates) who supply the standard configurable software product that covers:
  - an audit to establish assurance the suitability of the quality system used to develop the software (standard product)
  - review of critical algorithms, parameters, formulae
  - review of the data flow diagrams and resulting products for data integrity, security, accuracy, and reliability
  - review of the Part 11 elements contained in the standard product, such as logical security
  - writing and approval of an audit report
  - selection of the supplier of the standard product

Following the qualification activities used for a custom-built system performs the qualification of the configurable elements of the standard product. (Refer to *Custom-Built Systems* in this chapter.)

The qualification of the configurable elements of the standard product is performed according to normal software installation, operational and performance qualification practices. (Refer to *Software Installation Qualification, System Operational Qualification* and *System Performance Qualification* in this chapter.)

Change Control Must Be Rigorously Applied

The impact (criticality) of new versions on the standard product must be reviewed and appropriate action taken. After the new version has been installed, the new functionality must be tested and regression testing performed as part of the qualification activities.

If a manual system is being replaced, the two systems should be run in parallel for a time, as a part of the operational qualification.

User training should emphasize the importance of change control and the procedural controls to support the system

Data should be secured by physical and/or electronic means against willful or accidental damage. Stored data must be checked for accessibility, durability, and accuracy. If changes are proposed to the computer equipment or its programs, the checks mentioned above should be performed at a frequency appropriate to the storage medium being used.

Data must be protected by backing up at regular intervals. Back-up data must be stored for as long as is necessary to comply with regulatory and company requirements at a separate and secure location.

There should be adequate alternative arrangements made available for systems in case of a breakdown. The time required to bring the alternative arrangements into use should be related to their criticality and the possible urgency of the need to use them.

The procedural controls to be followed if the system fails or breaks down should be defined and qualified. Any failures and remedial actions taken must be recorded.

Procedural controls must be established to record and analyse errors, and to enable corrective action to be taken.

Part 11 Areas of Interest

This item is the same as for Standard Software Packages.

CUSTOM-BUILT SYSTEMS QUALIFICATION

Introduction

Custom-built systems produced for a customer specifically to meet a defined set of user requirements. The practices contained in this section apply to custom-built applications (e.g., PLC-based systems, the configurable elements within configurable software applications, any custom-built software applications, and any custom-built interfaces to other computer systems or databases) that are considered part of a computer system.

Objective

All computer systems performing regulated operations must be validated. To support computer systems validation, the custom-built system must be validated. The SLC must incorporate a formal development methodology with multiple inspections of the supplier's development and associate testing activities and qualification testing.

Practices

For these systems, the full life cycle must be followed for all elements of the system (hardware and software).

An audit of the application developer (internal or external) is required to examine their existing and established quality systems.

A validation plan must be developed, based on the results of the quality audit and on the complexity of the proposed custom-built system, which describes the development methodology, inspection, and qualification activities.

The qualification of the custom-built application must be performed according to normal software installation, operational and performance qualification practices. (Refer to *Software Installation Qualification, System Operational Qualification,* and *System Performance Qualification* in this chapter.) The following project activities depicted in Chapter 7 are applicable to custom-built systems:

- Define the scope of the system and its strategic objectives
- Perform a preliminary evaluation of technology
- Conduct a feasibility study
- Process description/operation, which include data flow diagrams and, if applicable, P&ID
- Complete the change control forms and approvals, including obtaining the approval of QA
- Prepare the system requirements specification deliverable
- Prepare and approve a Validation Plan
- Conduct a System (hardware, software, and process) risk study and Criticality/Complexity Study and document the results. Any new requirements that result from the risk study must be added to the requirements specification deliverable
- Approve the system requirements specification deliverable

To conduct a quality audit of the supplier of the custom-built product:

- Perform an audit to establish the suitability of the quality system used to develop the software (custom-built system)
- Review critical algorithms, parameters, and formulae
- Review the data flow diagrams and resulting product for data integrity, security, accuracy, and reliability
- Review the Part 11 elements contained in the custom-built system, such as logical security
- Write and approve the audit report
- Select the developer of the custom-built system
- Revisit the risk study
- Prepare the system specification deliverable
- After a technical review, approve the system specification deliverable
- Prepare the technical design specification deliverable
- Perform a technical review on the design specification
- Revisit the risk study
- Approve the technical design specification deliverable
- Build the custom-built application
- Conduct a physical and functional audit
- Start the development of the qualification protocols
- Conduct a Software Inspection
- Conduct Software Module Testing and Integration Testing
- Conduct an FAT and SAT
- Conduct an IQ
- Conduct an OQ
- Conduct a PQ
- Develop and approve the qualification reports and/or Validation Report
- Conduct a Process/Product Performance Qualification. *Note*: This activity is typically performed on manufacturing systems only
- Prepare a Product/Process Report
- Update the documentation regarding references to application software and their associated versions

The approved summary report and associated validation documentation should be submitted for retention at the appropriate record retention center.

Change control must be rigorously applied.

The impact (criticality) of new versions on the validity of the existing qualification documentation must be reviewed and appropriate action taken. After the new version has been installed, regression testing may be required.

If a manual system is being replaced, the two systems should be run in parallel for a time, as a part of the operational qualification.

User training should emphasize the importance of change control and the procedural controls to support the system.

Data should be secured by physical and/or electronic means against willful or accidental damage. Stored data must be checked for accessibility, durability, and accuracy. If changes are proposed to the computer equipment or its programs, the checks mentioned above should be performed at a frequency appropriate to the storage medium being used.

Data must be protected by backing up at regular intervals. Back-up data must be stored for as long as necessary to comply with regulatory and company requirements at a separate and secure location.

There should be adequate alternative arrangements made available for systems in case of a breakdown. The time required to bring the alternative arrangements into use should be related to their criticality and the possible urgency of the need to use them.

The procedural controls to be followed if the system fails or breaks down should be defined, and specified. Any failures and remedial actions must be recorded and evaluated.

Procedural controls must be established to record and analyse errors, and to enable corrective action be taken.

Part 11 Areas of Interest

This item is the same as for Standard Software Packages.

Chapter 11

SLC Documentation

REGULATORY GUIDANCE

The deliverables that are important in order to demonstrate that software has been validated include[1]:

- written design specification[2] deliverables describing what the software is intended to do and how it intends to do it
- written test plans based on the design specification deliverables, which include both structural and functional software analysis
- test results[3] and an evaluation[4] of how these results demonstrate that the predetermined design specification deliverables have been met

SLC DOCUMENTATION

Computer system documentation means records that relate to system operation and maintenance, from high-level design documents to end-user manuals. Sidebar 11–1 illustrates some typical SLC documentation. In the software-engineering context, computer systems documentation is regarded as software. All regulatory provisions applicable to software are also applicable to its documentation.

Computer system documents are generated during the project. These documents may be either printed material or electronic records, such as computer files, storage media, or film. The more documents that are written, the higher the cost of managing them due to the increasing difficulty of keeping the documents consistent with the computer system.

System documents must be available if needed for review. Obsolete information must be archived or destroyed in accordance with a written record retention plan. System documents can be grouped as depicted in Figure 11–1.

---

[1] FDA, *Guidance for the Industry: Computerized Systems Used In Clinical Trials*, April 1999.
[2] In this context 'design specification' refers to Table 11–1 for sample documentation.
[3] In this context 'test results' refers to execute qualification protocol(s).
[4] In this context 'evaluation' refers to qualification summary report(s).

Group	Sample SLC Documentation
Design and testing specifications	Functional requirements, hardware and software design, test plans, test specifications/scripts, risk analysis
Qualification records	IQ, OQ, and PQ
Operating documents	Written procedures, operating manuals, reference manual
Operating records	Processing log print-out, critical alarm print-out
Maintenance records and documents	Program or equipment modification requests, system status reports, repair orders, preventive maintenance, maintenance manuals, configuration management records
Performance records	Performance standard specifications, performance logs, performance summary reports
Other records	Service agreements, vendor qualification records, vendors audit reports

Figure 11–1. System Documents Grouping.

In addition to the above classifications, system documentation can be classified by the source of the system documents, i.e., the hardware/software supplier and the customer.

Supplier/integrator items:

- development specifications (e.g., detailed hardware and software requirements, overall software functionality, detailed test specifications)
- operational specifications and requirements (e.g., standard software packages and configurable software documentation)
- instruction manuals, training manuals
- diagrams, flow charts, and descriptions that define the software modules and their interactions
- test databases

Customer items:
For minimum requirements: refer to 'Regulatory guidance' in this chapter.

Refer to Sidebar 11–1 for a list of typical system documentation.

All systems-related documentation must be rigorously controlled to ensure that it is consistent with the system in operation, available for inspection by a regulatory agency in a timely fashion, and to ensure that the system can be properly operated and maintained.

The quality attributes, applicable to any type of document are that it is: unambiguous, appropriate, comprehensive, available, accessible, followed, and periodically reviewed. The content of the system documents must be consistent with the type of software developed, and

---

- *Validation plan*
- *Risk analysis*
- *System requirements specification deliverable(s)*
- *System specification deliverable(s)*
- *Technical design specification deliverable(s)*
- *Source code/configurable code/scripts*
- *Quality audit and technical review reports*
- *Testing plans and reports*

  – unit
  – integration
  – functional

- *FAT and SAT plans and results*
- *Qualification plans*
- *Validation/Qualification results and summaries*
- *Traceability analysis*
- *User manuals*

Sidebar 11–1. Typical SLC Documentation.

the criticality and complexity of the system. In addition, documentation must satisfy five basic elements; that they are: *written, appropriate, clear, verified,* and *approved.*

Computer system documents, which support system validation, are submitted for retention to the appropriate document control area and retained for a period at least as long as specified by the predicated regulations applicable to the system.

Computer system documents contain the following elements.

A title page that contains the following:

- a title that clearly indicates the nature of the document
- if the document is related to a specific software system, the title also references the software system and the version of the software system
- a revision level number for the document
- a date indicating the creation date of the final draft of the document (before the approval date)
- a unique document identifier
- clearly designated area(s) for review signatures including a statement that clearly indicates the significance of the signing (e.g., 'This document has been reviewed by ...'), an area for the signature(s), the typed name(s) of the signatories, and the date of each signing
- clearly designated area(s) for approval signatures including a statement that clearly indicates the significance of the signing (e.g., 'This document has been approved for execution by ...' or 'This validation report has been approved as complete by ...'), an area for the signature(s), the typed name(s) of the signatories, and the date of each signing

Each page of the document contains the following:

- a header containing an abbreviated indication of the title of the document and the document identifier number
- a footer that contains the revision number of the document and the issue date of the document (which must correlate with the title page)
- a page number or other unique page indicator, which also indicates the total number of pages (e.g., page x of y)
- test scripts and test results that are part of a formal inspections and testing activity. These items should contain appropriate information that clearly identifies them. This information may include: an identification number traceable to a requirement identification number, the revision number of the document, the test execution or review date, and the name and signature of the person who conducted the test or review

The completion of computer systems documentation should follow a set of predefined rules as follows:

- Always use water-insoluble ink of suitable darkness to ensure copying will accurately transcribe the numbers
- Pencils, white-out, or erasable pen must never be used
- Corrections should be made with a single line through the error followed by the initials and date of the person making the correction
- Blanks, lines, or spaces for recording data or comments that were not used when the document was completed should be crossed through or otherwise indicated as not applicable

Another, independent, person should verify calculations, algorithms, SLC documentation, source code, test script execution, test results, validation plans, and qualification protocols, and so on, as applicable.

Chapter 12

Relevant Procedural Controls

The integrity of the information managed by a computer system is protected by procedural controls, rather than the technology used to apply the controls. Procedural controls comprise any measures taken to provide appropriate instructions for each aspect of system development, operations, calibration, and validation. For computer systems, procedural controls address all aspects of software engineering, software quality assurance, and operation. In a regulated environment, these controls are fundamental to the operation of the computer system.

Procedures covering computer technologies may include:

- the operation and maintenance of the technology
- the authentication of users, access controls
- management of records and audit trails
- physical security disaster recovery
- protection of remote access points
- protection of external electronic communications
- personnel responsibilities, and the ongoing monitoring of computer systems

One of the most important procedures to be developed and followed is the administration and retention of electronic records. Part 11 requires the retention of electronic records in electronic form. For regulated systems in which electronic signatures are not implemented (hybrid systems) the electronic record requirements (Sub-Part B) in Part 11 are applicable, and the electronic records are maintained and retained in electronic form for the period established by the predicate rule.

The procedural controls applicable to computer systems must have QA controls equivalent to other regulated operations. If personnel do not have complete instructions or do not understand what they need to do, the outcome can be unexpected results or operation failures.

The quality attributes applicable to all procedural controls are that they are appropriate, clear, and consistent with related procedures, adhere to regulatory, industry, and company standards, and are approved, available, accessible, followed, controlled, and periodically reviewed. The SLC process must be consistent with the applicable computer systems validation procedural controls.

Appendix F contains some sample procedures that are mapped to Part 11.

Written procedural controls for computer systems should be available for the following activities:

- the development and maintenance of computer systems
- the operation of computer systems
- managing the access to the computer system and management of the records
  - retention and storage of data
  - prevention of unauthorized access to the data
  - loss management procedures to deauthorize lost, stolen, missing, or otherwise potentially compromised tokens, cards, and other devices

- system restart and the recover of data
- the management of electronic signatures

All procedural controls must be reviewed periodically in order to ensure that they remain accurate and effective. Where appropriate, a number of topics may be combined into a single written procedure. Wherever possible, procedures should be established at a level allowing them to be utilized across a number of implementation and maintenance projects and systems. When new procedures are needed, existing written procedures are reviewed to determine if they can be upgraded to include new requirements or processes. Redundant procedures must be avoided.

The project team and validation coordinator should determine which procedures are appropriate for a particular system development, validation, and/or maintenance activity.

Chapter 13

Change Management

INTRODUCTION

Adherence to change management practices for computer technologies provides a process by which a change to a computer system must be proposed, evaluated, approved or rejected, scheduled, tracked, and audited.

As required by the regulations, a procedural control for the implementations of changes must be defined and documented. A change may be requested due to:

> **Regulatory guidance**
>
> *Systems should be in place to control changes and evaluate the extent of revalidation that the changes would necessitate. The extent of revalidation will depend upon the change's nature, scope, and potential impact on a validated system and established operating conditions. Changes that cause the system to operate outside of previously validated operating limits would be particularly significant.*
>
> FDA, *Guidance for the Industry: Computerized Systems Used In Clinical Trials*, April 1999.

- perfective maintenance or to correct the system due to new requirements and/or functions
- adaptive maintenance or to correct the system due to a new environment that could include new hardware, new sensors or controlled devices, new operating systems, and/or new regulations
- corrective maintenance or to correct the system due to the detection of errors in the design, logic, or programming of the system
- preventive maintenance or to correct the system in order to improve future maintainability or reliability, or to provide a better basis for future enhancements

Change requests can result from both planned and unplanned events. Planned events can be scheduled, and can be the result of a planned expansion in business requirements. Emergency or unplanned events can result from software or hardware problems that prevent the application from performing its defined tasks.

Changes can be further classified as major or minor. The SLC activities to be performed are determined by the criticality and complexity of the system and/or the change. Responsible persons from QA, manufacturing, engineering, R&D, and/or IT perform the evaluation of the changes. A traceability analysis can help to determine the impact of the change(s).

The replacement of hardware requires a formal IQ and, if applicable, regression testing to verify the correct integration of the hardware with the system and application software. When the hardware is replaced by an equivalent item (e.g., same model, revision, and manufacturer) it is necessary to record the details of the replacement part.

Emergencies can arise that require an immediate or emergency change in order to prevent the disruption of business processes.

Changes and modifications to application software can introduce as many problems/errors as they solve unless they are rigorously controlled. Site changes and modifications to application software are particularly difficult to manage due to the lack of a formal, quality controlled, development environment.

CHANGE MANAGEMENT PROCESS

The following process is applicable for changes to any component of computer technology that performs functions on regulated products, including hardware, peripheral devices, system software and utilities, application software, documentation, and communications hardware and software. These components can be either purchased from an external supplier or developed in house by the company or its designee.

Change management procedural control must be in place when the validated system is released for use. These controls must address the identification and specification of the change.

Initiation is the process by which a change is requested. This process includes: documenting the change required; providing a rationale for the change and its anticipated impact on other areas of the system hardware and/or software; and forwarding the change request to the appropriate personnel (e.g., the management of the user area and the system support area).

Assessing the Risk, Criticality, and Impact of Change

The persons responsible for managing the change process evaluate the feasibility and impact of the proposed change in relation to their area of expertise and authority. Areas that should be included in the evaluation process are, as applicable: QA, regulatory compliance, area management, the system owner/sponsor, the end-user, and technology professionals (e.g., IT, engineering).

Their conclusions are then forwarded to the designated approver(s).

Specification of the Testing Requirements and Acceptance Criteria, and Implementation of Change after Authorization

The approval or rejection of the proposed change by the identified responsible person(s) must be documented. Areas to be included in the approval process are, as applicable: QA; regulatory compliance; area management; the system owner/sponsor; the end-user; and technology professionals (e.g., IT, engineering). The approver(s) are responsible for evaluating the recommendations of the reviewers, making a decision on whether to proceed with the proposed change, and initiating the implementation of the requested change. The appropriate personnel must perform the change required in accordance with the company policies, documented practices and/or system maintenance methodology.

Following the successful testing of the change, standard operating procedures are used to:

1. control the transfer of the change into the operational environment
2. document a 'backout/rollback plan'
3. verify functionality
4. document the final change
5. notify any impacted personnel
6. train any impacted personnel

During the move into the operational environment, any deviations and/or errors are identified that cannot be corrected, the documented backout/rollback procedure must be followed to restore the system to its premodified state.

Performing Applicable Regression Testing

All changes must be tested in the appropriate environment and documented according to defined procedures that include:

- a detailed plan for testing the change
- the development of test script(s) for the new functionality, requirements, and/or environment
- the execution of the test(s)
- the analysis and recording of the test results

Reviewing the Change(s) with an Independent Reviewer

Changes to computer systems must be reviewed in order to verify its compliance with the change management process, to ensure that the appropriate procedures have been followed, and that the validated status of the system has been maintained. The results of the audit are recorded and saved as part of the change control file for the computer system.

Updating the system and user documentation must always reflect the implemented change(s).

Emergency change procedurals control must consider the specific requirements of each computer system.

Provisions must be established for the management of an 'emergency change,' including expeditious documentation modifications.

Chapter 14

Training

REGULATORY GUIDANCE[1]

Determination that persons who develop, maintain, or use electronic record/electronic signature systems have the education, training, and experience to perform their assigned tasks.

TRAINING IN THE REGULATED INDUSTRY

Personnel operating, maintaining, and programming computer systems should have adequate training and experience to perform the assigned duties. One aspect to consider is the extent of operation and system management. Training should include not only system operation but also cover the significance of system faults (bugs), regulatory requirements, system changes, security procedures, manual operation of the system, and documentation of system errors. The organization must record the training of computer system related personnel.

Training records are managed and retained by either the department providing the training or by another designated department. The project leader for new systems, or the system administrator for existing systems should coordinate the training. The system administrator may maintain a copy of the training records throughout the life of the computer system. Depending on the system, these records may include:

- a description of the supplied training
- a description of in-house training
- copies of training manuals/materials

Any training conducted online must be performed in a controlled environment to ensure that production systems and data are not adversely impacted. Appropriate security measures must be taken to ensure that the training is isolated from the production environment.

A linkage should be provided between training programs and the change management process so that training requirements can be continually evaluated, adjusted, and maintained in alignment with the current state of the operational computerized system.

---

[1] Department of Health and Human Services, Food and Administration, *21 CFR Part 11, Electronic Records; Electronic Signatures*, Federal Register 62 (54), 13430–13466, March 20, 1997.

Training manuals, handouts, personnel proficiency tests, and training procedures supplied as part of the training provided, are an integral part of the system's documentation. If these are available electronically, they can be referenced.

Training plans should be designed so that they are suitable for the intended audience. As a result, there may be different courses for managers, supervisors, operations personnel, system administrators, and other types of users.

Chapter 15

Security

REGULATORY GUIDANCE[1]

Access to electronic records should be restricted and monitored by the system's software through its log-on requirements, security procedures, and audit trail records. The electronic records must not be altered, browsed, queried, or reported by external software applications that do not gain entry through the protective system software. In addition to the logical security built into the system, physical security must be provided to ensure that access to computer systems and, consequently, to electronic records is prevented for unauthorized personnel.

INTRODUCTION

To create a trusted digital environment within an enterprise, it must be ensured that both the application and network components are truly secure. The FDA addresses the subject of security for computer systems in the CGMP regulations and their associated policy guidelines. Specifically, 21 CFR Part 211.68(b) and a recent guideline[2] require that appropriate controls are established over application and network components to ensure that only authorized personnel can make changes to master

production, control and/or other required records. CPG 7132a.07, Inputs/Outputs Checking, contains specific requirements to establish the necessary controls for records.

The first fundamental element of Part 11 is that it requires organizations to store regulated electronic data in its electronic form, rather than keeping paper-based printouts of the data on file, as had been the long-term practice in organizations performing regulated operations. If information is not recorded on durable media, it cannot be retrieved for future use. When 'retrievability' is a requirement, then the procedural and technological controls contained in Part 11 are essential to ensure data integrity. Only regulated electronic records that meet Part 11 may be used to satisfy a predicate rule.

The second fundamental element in Part 11 is the high level of security required to protect regulated electronic records. The requirements for application and network component security can be found in Table 15–1. The controls that are implemented resulting from the security related requirements are intended to build a trusted digital environment. The attributes relevant to a trusted digital environment are:

• private (secure information)

---

[1] FDA, *Guidance for Industry: Computerized Systems Used In Clinical Trials*, April 1999.
[2] FDA, *Guidance for Industry: Computerized Systems Used in Clinical Trials*, April 1999.

Table 15–1. Part 11 security related requirements/controls.

Part 11	Description	PDA/ISPE Technological Controls[3]
11.10(c)	Protection of records	The system should be able to maintain electronic data over periods of many years regardless of upgrades to the software and operating system.
11.10(d) 11.10(d)	Access controls Authentication	The system should restrict access in accordance with preconfigured rules that can be maintained. Any changes to the rules should be recorded.
11.10(e) 11.10(e)	Audit trail controls Computer systems time controls	The system should be capable of recording all electronic record creation, update, and deletion perations. This record should be secure from subsequent unauthorized alteration.
11.10(g)	Authority checks	The system should restrict the use of system functions and features in accordance with preconfigured rules that can be maintained. Any changes to the rules should be recorded.
11.10(h)	Device checks	Where pharmaceutical organizations require that certain devices act as sources of data or commands, the system should enforce this requirement.
11.30	Technical controls to open systems	Not covered by the PDA/ISPE publication.
11.70	Signature/record linking	The system must provide a method for linking electronic signatures, where used, to their respective electronic records, in a way that prevents the signature from being removed, copied, or changed in order to falsify that or any other record.
11.100(a)	Uniqueness of electronic signatures	The system should enforce uniqueness, prevent the relocation of electronic signatures, and prevent the deletion of information relating to the electronic signature once it has been used.
11.300	Electronic signatures security	The system should be able to identify changes to electronic records in order to detect invalid or altered records.

[3] ISPE and PDA, *Good Practice and Compliance for Electronic Records and Electronic Signatures, Part 2, Complying with 21 CFR Part 11, Electronic Records and Electronic Signatures,* September 1, 2000.

- authentic (proof of identity)
- reliable (information integrity)
- nonrepudiatable (undeniable proof of sender/receiver)

Another area of concern for the FDA relates to records sent over public networks, dial-up connections, or public phone lines, accessed through external/internal web servers or by database servers. A number of security issues need to be considered in order to keep these records trustworthy. Intranets may face similar security issues if remote users connect to the central network resources through a local link to an Internet, even when password-protected access for the users is provided to just a small portion of a private network. These systems are considered by the agency as Open Systems (Part 11.30)[4]. For example, the Internet provides a convenient medium to connect to other networks, but it does not provide reliable security features, such as entity authentication, or protection from hostile users or software.

In the USA, the Social Security Administration's policy[5] is to integrate security into the SLC for each application for the following reasons:

- **It is more effective**. Meaningful security is easier to achieve when security issues are considered as a part of the routine development process, and security safeguards are integrated into the system during its design
- **It is less expensive**. To retrofit security is generally more expensive than to integrate it into an application during development
- **It is less obtrusive**. When security safeguards are integral to a system, they are usually easier to use and less visible to the user

Figure 15–1 depicts the types of security issues to consider in the Internet/intranet world. A vital factor necessary to achieve trustworthy records is to implement electronically based solutions

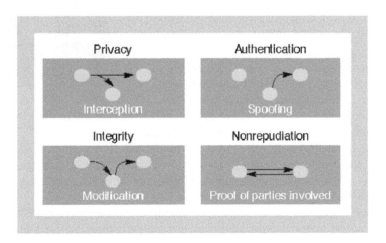

Figure 15–1. Security issues to consider.[6]

---

[4] FDA, *Human Drug CGMP Notes*, June 1999 (Vol. 7, No. 2).
[5] US Social Security Administration, *Systems Security Handbook*, Release 5.0, Chapter 6.
[6] D. Coclin (Dean.Coclin@baltimore.com), *Public Key Technology Overview*, Courtesy of Baltimore Technologies (www.baltimore.com).

to strengthen the security. Trustworthy records must also be considered as part of the requirements for record retention, archiving, and retrieval.

System security is also a concern for computer technology suppliers and contract developer's sites. The integrity of the computer environments that are used to maintain the application software has to be verified when auditing suppliers and/or developers.

An assurance that the physical hardware, software, and the regulated electronic records are maintained in a secure environment is critical to the validated status of a computer system, particularly if it is an enterprise level system. Security must be instituted at several levels. Procedural controls must govern the physical access to computer systems (*physical security*). The access to individual computer system platforms is controlled by network specific security procedures (*network security*). Finally, application level security and associated authority checks control access to the computer system applications (*applications security*).

PHYSICAL SECURITY

The following items are the key practices that are applicable for the security of the physical environment of a computer system.

1. Limit physical computer access to authorized individuals.
2. The power sources for computer equipment must be controlled in order to prevent an accidental uncontrolled loss of power, and fluctuations in voltage levels, which could potentially damage equipment, software files and data files.
3. The power supplies for all computer equipment must meet both the manufacturers requirements and safety national requirements for wiring, grounding, and radiation.
4. Critical environmental parameters that include, but are not limited to, relative humidity and temperature, must be maintained within the computer equipment manufacturer's requirements. In cases where there are a number of requirements, the most stringent should be utilized.
5. Critical environmental parameters must be monitored on a regular periodic schedule in order to ensure that they continue to meet the manufacturer's requirements.
6. The facility where the computer equipment resides must be equipped with the appropriate fire-extinguishing systems or equipment.
7. The computer operations support, personnel facilities maintenance, and all other personnel who need access to the physical environment must be responsible for knowing where and how to obtain the applicable operational procedures, understanding their contents, and for accurately performing their execution.
8. The use of authentication checks at physical access points in order to establish that the machine-readable codes on tokens and PINs are assigned to the same individual. These checks must be performed on the controlled access location(s) where the computer hardware is resident.
9. The use of 'time-outs' for under-utilized terminals in order to prevent unauthorized use while unattended.

NETWORK SECURITY

The following items are the key practices that are applicable for the security of the network environment.

1. Network resources have a qualified authentication mechanism for controlling access (Part 11.10(d)). Table 15–2 shows the authentication requirements and the methods of implementation. These authentication requirements and methods of implementation are also applicable to applications security.

Table 15–2. Authentication.

Requirement	Implementation
The following features must be implemented: • Automatic logoff, • Unique user identification. In addition, at least one of the other listed implementation features must be a procedure to corroborate that an entity is who it claims to be.	• automatic logoff • biometrics • password • PIN • telephone callback • token • unique user identification

Access control decision function are defined using access right lists,[7] such as Access Control Lists (ACLs), and these allow the allocation of use, read, write, execute, delete, or create privileges. Access controls enable a system to be designed in such way that a supervisor, for example, will be able to access information on a group of employees, without everyone else on the network having access to this information.

In the context of Part 11, access controls are an element of the authority checking requirements.

2. The process for setting up user access to a network is the responsibility of the appropriate network security administration personnel. The technical preparation, education, and training for personnel performing administration tasks are fundamental. The determination of what is required and the provision of documented evidence of technical preparation, and the education/personnel training, are key regulatory requirements (Part 11.10(i)).

---

[7] In Windows® 2000, an access control list (ACL) is a list of security protections that apply to an entire object, a set of the object's properties, or an individual property of an object. Each Active Directory object has two associated ACLs: The *discretionary access control list* (DACL) is a list of user accounts, groups, and computers that are allowed (or denied) access to the object. A DACL consists of a list of access control entries (ACEs), where each ACE lists the permissions granted or denied to the users, groups, or computers listed in the DACL. An ACE contains a security identification with permission, such as Read access, Write access, or Full Control access. The *system access control list* (SACL) defines which events (such as file access) are audited for a user or group.

3. Procedural controls that specify the manner in which network security is administered must be established and documented. Network users must be trained in the policies, practices, and procedures concerning network security.

4. The management of network user accounts is a key procedural control. This process includes the request for the addition, modification, and removal of access privileges (Part 11.10(g)). The request must be approved by the appropriate manager, documented, and submitted to the network security administration for implementation.

5. There must be a procedure for granting controlled temporary network access for personnel ((Part 11.10(d)).

6. In the event that a user leaves the company, there must be a process for notifying the appropriate security administration as soon as the employee departs.

7. Provisions must be made for the regular monitoring of access to the network. There must be an escalation procedure for defining the actions to be taken if unauthorized access to the network is discovered.

8. A documented record of security administration activities must be retained.

9. Procedures must be established to control remote access to the network. Systems that have connections to telephone communications through modems should have strict access controls and restrictions. Access restrictions on these systems should be designed to prevent unauthorized access or change. One possible method for controlling remote access is telephone callback.

10. The use of time-stamped audit trails (Part 11.10(c) and (e)) to record changes, to record all write-to-file operations, and to independently record the date and time of any network system administrator actions or data entries/changes.

11. Unauthorized modification of the system clock must be prevented (Part 11.10(d)).

APPLICATIONS SECURITY

The following items are key practices that are applicable for the security of applications. These practices are applicable to both networked and stand alone applications.

1. All applications must have a qualified authentication mechanism to control access (Part 11.10(d)).

2. Software 'virus checking' must take place periodically for the protection of the application and data.

3. For each application, procedural controls must be established that specify the manner in which application security is administered.

4. The process for setting up access to applications must be defined and executed by the appropriate, application specific, security administration personnel. The technical preparation, education and training for personnel performing administration tasks is fundamental. The determination of what is required and the documented evidence of the technical preparation, education and training of personnel is a key regulatory requirement (Part 11.10(i)).

5. The management of the user application accounts is a key procedural control. This process includes the request for the addition, modification, and removal of application access privileges (Part 11.10(g)). The request must be approved by the appropriate manager, documented, and submitted to the application security administration for implementation.

6. There must be a procedure for granting controlled temporary application access for personnel. As in the regular access, temporary access must follow security best practices (Part 11.10(d)).
7. In the event that a user leaves the company, there must be a process for notifying the appropriate security administration as soon as the employee departs.
8. There must be an escalation procedure for defining the actions to be taken if unauthorized access to the application is discovered.
9. A documented record of security administration activities must be retained.
10. Procedures must be established to control remote modem access to applications.
11. If data input or instructions can only come from specific input devices (e.g. instruments, terminals), the system must check the source of the input and the use of the correct device should be verified by the operator (Part 11.10(h)).
12. The use of time-stamped audit trails (Part 11.10(c) and (e) to record changes, to record all write-to-file operations, and to independently record the date and time of any application specific operator actions or entries.
13. The use of time-stamped audit trails (Part 11.10(c) and (e) to keep track of any record modifications carried out by the database administrator.
14. Use operational checks to enforce sequencing (Part 11.10(f)). Refer to Chapter 20 for additional information.
15. The use of authority checks (Part 11.10(g)), when applicable, to determine if the identified individual has been authorized to use the system, and has the necessary access rights, operate a device, or to perform the operation required.
16. Unauthorized modification of the system clock must be prevented (Part 11.10(d)).
17. A 'shared' computer system is a system used by various applications. Security procedures should have been established regarding limiting access to confidential information. In addition, access to data should be limited so that inadvertent or unauthorized changes in data do not occur.

OTHER KEY SECURITY ELEMENTS

Authentication

Three types of authentication are discussed in relation to Part 11: person authentication, records/messages authentication, and entity authentication.

There are three *person authentication* methods: PIN and static passwords, PIN and dynamic passwords, and biometric devices. Typically, the person authentication process starts when the person enters a PIN into the system, and then his/her identity is authenticated by providing a second piece of information that is known, or can be produced only by the person (e.g., a password).

The most common methods for providing a strong authentication comprise of automatic password generators (e.g., tokens) and smartcards. Tokens and smartcards store information about the person and require the use of a reader device. To protect against theft, the person must enter a password or PIN before the information in the token or smartcard can be accessed.

Person Authentication

- *PINs and static passwords*
  Static passwords are still the simplest, most cost-effective and widely used authentication

mechanism. Although there are stronger mechanisms than passwords (e.g. one-time passwords, challenge-response mechanisms, cryptographic mechanisms, and biometric mechanisms) to achieve person authentication, passwords are the most common mechanism, and often provide adequate protection for computer resources.

Most systems today offer at least basic password management features. For the password management features to be consistent with Part 11.300, they should include:

- Password aging and expiration — this allows the password to have a lifetime, after which it will expire and must be changed
- Password history — this checks the new password to ensure that it has not been reused during a specified amount of time or the specified number of password changes
- Account locking — this provides the automatic locking of an account if a person fails to log in to the system after a specified number of attempts
- Password complexity verification — this performs a complexity check on the password to ensure that it is complex enough to provide reasonable protection against intruders who might want to break into the system by guessing the password
- Passwords file security — Passwords file shall be protected or otherwise secured so that the password cannot be read by ordinary means. If the password file is not protected by strong access controls, the file can be downloaded. Password files are often protected with one-way encryption so that a plain-text password is not available to system administrators[8] or hackers (if they successfully bypass access controls). Even if the file is encrypted, brute force can be used to learn passwords if the file is downloaded (e.g., by encrypting English words and comparing them to the file)

- *PIN and dynamic passwords*
  Passwords and PINs can provide a stronger means of authentication when combined with other authentication techniques. Typically, the combination involves something the person knows (e.g., passwords, PINs) and something that the person possesses (e.g. tokens). Tokens are a software or hardware mechanism that provides the person with a second piece of authenticating information. Included within this group of mechanisms are:
  - One-time password generators — As the name implies, one-time passwords are similar to traditional passwords since they are used in conjunction with a PIN, but they are limited to one-time use. The advantage of this technique is that it prevents the reuse of a compromised password. Commonly, a small hand-held device the size of a credit card is synchronized with the target system's authentication scheme and displays a one-time password that periodically changes. To access the target system, the person enters an assigned PIN and password followed by the one-time password currently displayed on the hand-held device. This method of authentication provides additional security since the person must possess knowledge of some personal information as well as the authentication token.
  - Challenge-response schemes — These look like one-time password generators and use a similar synchronization mechanism; however, additional user actions are required for authentication. It involves a challenge/response exchange with a new key being used at each login.

---

[8] One-way encryption algorithms only provide for the encryption of data. The resulting ciphertext cannot be decrypted. When passwords are entered into the system, they are one-way encrypted, and the result is compared with the stored ciphertext.

- • SSL 3.0 encrypts data between the server and browser. In addition it supports client authentication for Access Control.

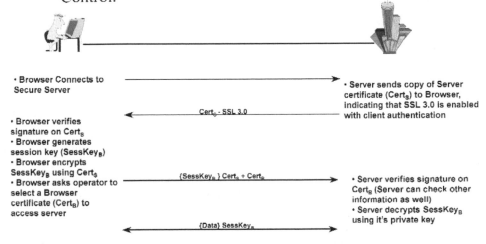

• Browser Connects to Secure Server

• Server sends copy of Server certificate (Cert$_s$) to Browser, indicating that SSL 3.0 is enabled with client authentication

Cert$_o$ - SSL 3.0

• Browser verifies signature on Cert$_s$
• Browser generates session key (SessKey$_B$)
• Browser encrypts SessKey$_B$ using Cert$_s$
• Browser asks operator to select a Browser certificate (Cert$_B$) to access server

{SessKey$_B$} Cert$_s$ + Cert$_B$

• Server verifies signature on Cert$_B$ (Server can check other information as well)
• Server decrypts SessKey$_B$ using it's private key

{Data} SessKey$_B$

• Browser and Server use SessKey$_B$ to encrypt all data exchanged over the Internet

Figure 15–2. SSL 3.0 Protocol.[9]

- – Certificate-based authentication — Certificate-based authentication is an increasingly popular technique for strong authentication. An entity (see the definition later in this chapter) wishing to be authenticated presents a digital certificate to the security system. If the certified party is still trusted by the organization, then the server trusts that the entity is who it claims to be. One benefit of certificate-based authentication is that the entity does not have to have an established a relationship with, for example, a server prior to being authenticated. This makes certificate-based authentication an excellent solution for on-line commerce.

    SSL 3.0 (Figure 15–2) protocol uses x.509v3 compliant digital certificates for authentication. The most common usage of SSL is to authenticate a server to a person, but it can also be used to authenticate a person to a server. By authenticating a server, the person is assured of communicating with the correct site. SSL also signs the information transmitted so that the person is assured that the information has not been changed in transit.

    A similar mechanism to certificate-based authentication is the authentication of email transport. The authentication of email transport can be checked against the identity of the recipient. The authentication of email users ensures that there is no fraudulent user access to the system.

- – Biometric Devices — Biometric mechanisms tend to be best suited for the enforcement of physical access, rather than for electronic access. That is, they work well where a human guard (or proximity sensor, etc.) might be used, but they are not generally used

---

[9] D. Coclin (Dean.Coclin@baltimore.com), *Public Key Technology Overview*, Courtesy of Baltimore Technologies (www.baltimore.com).

for electronic access to a Web site. Consider, as an example, that the biometric device in question (e.g., your fingerprint) is to be used for access to a Web site. An attacker only has to capture your fingerprint data once, and they can then replay it while masquerading as you. The following Internet address provides a list of biometric devices: http://www.icsa.net/ services/consortia/cbdc/certified_products.shtml.

Records/Messages Authentication

Records must remain complete during transmission. An *authentic record* is one that is proven to be what it purports to be and to have been created or sent by the person who purports to have created and sent it. *Records/messages* authentication provides the means to evaluate the integrity of the data/message. It involves technical and procedural controls to detect unauthorized modifications to data/messages.

Unauthorized manipulation of records, audit trail records, and replay of transmissions must be reliably identified as errors. The recipient can use three approaches to authenticate data/messages.

- encrypting records/messages using the recipients' public key
- signing the records/messages
- employing a session key

The details of the above technologies are outside of the scope of this book. Additional information on records/message authentication can be found in Chapter 22.

Entity Authentication

An entity can be a person, server, organization, or an account. One example of *entity authentication* is Kerberos (see Internet address http://www.isi.edu/gost/info/Kerberos/). Kerberos is an industry-standard authentication system suitable for distributed computing over a public network. Implemented in Windows® 2000, it uses public-key encryption and hashing algorithms. Kerberos requires a Certificate Server that acts as a certification authority and manages keys and digital certificates.[10]

This server maintains a database of secret keys for each entity, authenticates the identity of an entity that wishes to access network resources, and generates session keys when two entities wish to communicate securely.

The details of the above technologies are outside of the scope of this book.

Protection of Records and Audit Trails

Protection of records

All operations with records must be performed in a secure environment. This requirement is applicable to:

---

[10] D. Coclin (Dean.Coclin@baltimore.com), *Public Key Technology Overview*, Courtesy of Baltimore Technologies (www.baltimore.com).

- records with/without signatures
- records in transit

Organizations can use access controls, encryption (e.g., end-to-end data encryption) and/or hashing to protect the integrity, authentication and privacy of records and/or to prevent the loss of records. Records must be protected, for example, when confidential patient data for specialist institutions is sent to distant places. Access controls were discussed previously in this chapter.

Encryption can be used to protect records in transit between networks, including the Internet. Tools such as network diagnostics or protocols analyzer can easily see information as it is transmitted. This can be prevented by, for example, using SSLs, TLS, or VPNs provide the necessary management of critical data across networks. These solutions use encryption, digital signatures, and/or digital certificates to ensure data privacy, identification of the originator of messages, and the verification of message integrity.

End-to-end data encryption using PKI technology is implemented by encrypting messages/records using the recipient's public key, thus rendering messages/records unreadable to anyone except the recipient. To encrypt messages/records the CA issues distinctive digital signing certificates containing the public key assigned to the recipient. Messages/records can only be decrypted using the recipient's private key. The combination of hashing and access controls provides another opportunity to protect sensitive records.

An alternative method for sending records/messages via the Internet is using Secure/Multipurpose Internet Mail Extensions (S/MIME). It is the standard for secure e-mail within PKI. It enables secure communication between two or more parties by ensuring the integrity and authenticity of the message given a digital signature, and the confidentiality of the message if the message is encrypted.

Scheduled backups, established restoration procedures and off-site storage of media are best practices that support recovery of records. The metadata associated with each record must also be saved.

The protection of records by archiving must include taking into consideration the rate of degradation of the electronic images, and the availability of the devices and software needed to access the records.

Audit Trails

Audit trails refer to a journal that records modifications to electronic records. The persons or automated processes operating on the user's behalf may perform these modifications. The audit-trail mechanism provides the capability to reconstruct the data prior to a modification, and therefore must not obscure the previously recorded data. The tracking mechanism must include a computer-generated time stamp that indicates the time of the entry. Audit trails are computer generated and can be part of the record that has been modified or a standalone record linked to the modified record.

The date and time of the audit trail must be synchronized to a trusted date and time service. More information on this is provided in the next chapter.

One of the key controls for audit trails is the linking of the electronic record with the audit trail. It must not be possible to modify audit trails; access rights to audit trial information must be limited to print and/or read only. The combination of authentication, digital certificates, encryption and ACLs, provides the technical mechanisms needed to control the access to audit trail files.

Chapter 16

Source Code

REGULATORY GUIDANCE

We regard source code and its supporting documentation for application programs used in drug process control to be part of master production and control records, within the meaning of 21 CFR Parts 210 and 211. Accordingly, those sections of the current good manufacturing practice regulations, which pertain to master production and control records, will be applied to source code.[1]

When the end user cannot directly review the program source code or development documentation (e.g., for most commercial off-the-shelf software, and for some contracted software) more extensive functional testing might be warranted than when such documentation is available to the user. More extensive functional testing might also be warranted where general experience with a program is limited, or the software performance is highly significant to data/record integrity and authenticity.[2]

INTRODUCTION

Source code is characterized as the human readable version of the list of instructions (program) that cause a computer to perform a task. Source code in this context must be viewed with a wide meaning since it includes, for example, the program listings for custom-built applications; the configurable elements or scripts for configurable applications; and the critical algorithms, parameters, and macro  listings for off-the-shelf applications. The source code typically includes, as part of the code, comments that describe the subroutines or modules of the language used.

The importance of the source code comes from the necessity to maintain the program and to evaluate scripts, macros, or process specific parameters, and evaluating the impact of any change on the rest of the program.

Up to the late 1980s, 80% of computer systems were specifically developed (i.e., custom/bespoke applications) as requested by the drug manufacturer and the source code was a deliverable. By the early 1990s, the increased use of computers led to an increase in the use of Standard Software Packages (see Chapter 10). Today, 80% of the software utilized to supervise and control manufacturing processes is of the Standard Software Packages type and only 20%

---

[1] FDA, Compliance Policy Guide 7132a.15, *Source Code for Process Control Application Programs*, April 1987.
[2] FDA, draft of *Guidance for the Industry: 21 CFR Part 11; Electronic Records; Electronic Signatures: Validation*, August 2001.

is specifically developed (i.e., custom/bespoke applications). Many of the specifically developed programs can be found, for example, in programmable logic controllers. For 'canned configurable' software applications, the source code remains the proprietary information of the developer, so the likelihood of acquiring it is very low.

Code inspections, reviews, and audits (see Chapter 9) consist of the smallest unit of software designed to have a defined purpose during software development. These debugging activities are necessary to verify that the code conforms to the software requirements and the software development procedures. When testing the programs the test cases are generated based on the internal code structure and include a 100% path coverage. Code inspections and reviews are long and laborious processes performed by the programmer and his/her peers; it is very difficult to perform code inspections and reviews after the system has been installed.

Any serious software developer will follow a formal software development methodology that includes the documentation necessary to support the software at each development step. A quality audit of the software developer will determine the adequacy of testing performed and whether the tests plan, procedures, and results are properly documented. The quality audit must include an inspection of the unit testing process.

In 1996,[3] the FDA provided an alternative to the source code review. When the source code and the system specification cannot be obtained from the supplier, then *'black box testing must be performed to confirm that the software meets the user's needs and its intended uses.'* One way to make the source code available for the standard products of configurable applications and for some contracted software is by establishing a third-party escrow account. This account enables organizations to ensure that application software source code can be made available for inspection, if requested by a regulatory agency. If an escrow account is implemented, organizations must adhere to the confidentiality, copyright, and other legally binding agreements concerning the software source code.

Before deciding on an escrow account, the legal department of the pharmaceutical company must clearly communicate to the computer system supplier the source code access requirements expected. Organizations should not require source code availability for the following software: operating systems and system software, instruments and controllers, and standard software packages.

When the criticality and complexity (Refer to Appendix D) of the computer system requires that standard product source code be made available, the legal department of the pharmaceutical company should negotiate one of the following written agreements with the software developer.

- to deliver the source code to the organization
- to place the source code into a third-party escrow account accessible to the organization
- that the source code will be made available for inspection by a representative of a regulatory agency

The agreement should also cover the updating of the source code held in escrow, and the exact versions of which software applications must be held in escrow.

In any case, the software supplier or the developer must follow a documented change management procedure, so that changes to source code are specified and documented.

---

[3] Department of Health and Human Services, Food and Administration, *21 CFR Parts 808, 812, and 820 Medical Devices; Current Good Manufacturing Practice (CGMP); Final Rule,* Federal Register, 61 (195), 52602–52662, October 7, 1996, comment paragraph 136.

Chapter 17

Hardware/Software Suppliers Qualification

It is the practice of organizations performing operations in a regulated environment to work proactively with computer technology suppliers and contract developers, to ensure that the system components conform to the quality requirements defined by each organization.

When purchasing system software, standard instruments, microcontrollers, smart instruments, and standard software packages, it is the practice of organizations to initially rely on the history of the supplier and the industry operating experience as assurance of the supplier capability of program performance suitability.

The history of the supplier, and the operating experience within the industry, can be regarded as an assurance that the supplier is capable of providing a program that performs

> **Regulatory guidance**
>
> *Systems should be in place to control changes and evaluate the extent of revalidation that the changes would necessitate. The extent of revalidation will depend upon the change's nature, scope, and potential impact on a validated system and established operating conditions. Changes that cause the system to operate outside of previously validated operating limits would be particularly significant.*
>
> FDA, *Guidance for the Industry: Computerized Systems Used In Clinical Trials*, April 1999.

in a suitable manner. However, the history of the supplier and the industry's operating experience alone are not sufficient proof of a program's performance. The end-user is responsible for ensuring that the program is suitable for its intended purpose when used in a regulatory environment. Refer to Chapter 7 for additional validation activities.

When purchasing configurable software, it is often the practice to evaluate the criticality of the application to the regulatory operation, in order to determine if a supplier evaluation is warranted and feasible. If a supplier evaluation is not warranted or feasible, the rationale for this must be documented and retained with the system validation documentation. As in the previous case, the end user is responsible for ensuring that the program is suitable for its intended purpose when used in a regulatory environment.

For custom-built application software, it is normal practice for organizations to evaluate the developer's quality system, including his procedural controls and experience, in order to ensure that the system will conform to the quality requirements defined by the organization.

The purpose of performing a hardware/software supplier qualification is to assess the computing environments and technology products used to develop regulated applications.

These qualifications enable an objective evaluation of the supplier to be performed and documented, including any findings, and to provide a baseline for computer system validation.

The evaluation is accomplished through the inspection of:

- practices used for the development, testing, maintenance, and use of the computing technology
- documented evidence that supports the integrity of the computing environment and the technology products developed
- historical data on the available operating experiences

During the qualification of computer technology suppliers and developers, a site visit and inspection of the computing environment and related technology products is conducted. Information is collected through formal inspection of the defined practices, verification of supporting evidence, and interviews with key personnel. Information collected is compared to the defined acceptance criteria and verified before completion of the on-site visit.

Items to be considered are:

(a) **Management**
- A defined organization is in place for the management of the project or contract
- A quality assurance function is represented
- The roles and responsibilities are clearly defined

(b) **Documentation**
- Developer's system specification exists and is linked with the organization's process/operational requirements
- Detailed hardware and/or software design specifications are documented
- Hardware and/or software verification plans are documented
- Test plans, procedures, scripts/specifications, input, and output data, and test reports are available for inspection/audit
- Verification of hardware and/or software functionality is documented
- Verification of adherence to standards or development practices is documented
- The source code is available, if applicable. Refer to Chapter 16
- Site acceptance test plan and results are documented (if applicable)
- User and technical support manuals and documentation are available

(c) **Standards and procedures**
- Documented evidence is available that the hardware and/or software development process is conducted under the auspices of documented standards, methodologies, conventions, or procedures.

(d) **Configuration management**
- A documented system is available and followed for configuration management. Configuration management applies to: specifications, design and test documentation; test plans and procedures; each item of hardware; development of software at each incremental stage and for each constituent part/program and module; and related documentation (e.g., manuals and reports).
- Procedures for the security, archiving, and recovery of software products are clearly defined.

(e) **Change management**
- An existing and documented management of change process is followed during the development and testing stages, and extending to the on-site installation and commissioning activities, if required under the contract. This includes mechanisms for requesting, submitting, evaluating, the impact, approving, implementing, testing, releasing, and documenting the change. The traceability analysis can be used to quickly identify areas impacted by the change.
- The policies and procedures for the ongoing support and management of current and obsolete products are clearly defined.

(f) **Personnel qualifications**
- Documented evidence is available that establishes that the personnel designing, programming, and maintaining the computer system have adequate training and experience.

After visiting and inspecting the supplier, the information obtained is analyzed against the acceptance criteria, current company requirements for the validation of computer systems, and the regulatory expectations for computing environments. A report is produced that documents the audit process, provides information on the current state of validation of the supplier and what may be needed for compliance and, based on the audit findings, provides a recommendation on whether the supplier can be used.

Any corrective actions that are required must be defined and implemented in order to ensure that a validated state is achieved. The corrective actions must be scheduled and follow-up evaluations planned for monitoring and tracking their progress.

Chapter 18

Maintaining the State of Validation

Once the system has been released for operation, the computer system maintenance activities take over. The maintenance activities must be governed by the same procedures followed during the development period.

The validated status of computer systems performing regulated operations is subject to threat from changes in its operating environment, which may be either known or unknown. Adherence to security, operational management, business continuity, change management, periodic review, and decommissioning procedures and policies, provides a high degree of assurance that the system is being maintained in a validated state. It is essential that organizations have procedures in place to minimize the risk of computer systems performing regulated operations in a nonvalidated state.

The maintenance of computer systems becomes an essential issue, particularly when a new version of the supplier-supplied standard software is updated. A change control procedure must be implemented whereby changes in the software and computer hardware may be evaluated, approved, and installed.

When necessary, an additional analysis may be needed to evaluate the changes (e.g. an impact analysis) on a computer system. The change control procedure should allow for both planned and emergency changes to the system, and must include a provision for the updating of the appropriate system documentation, including any procedures that may be applicable. Records of changes to the system must be kept for the same period as any other regulated production document.

Table 18–1[1] summarizes the periods and events applicable during the operational life of computer systems and the associated key practices.

SECURITY

Security is a key component for maintaining the trustworthiness of a computer system and associated records. Security is an ongoing element to consider, subject to continuous improvement. The security measurements that are implemented on a computer system may be obsolete after several years.

---

[1] G. J. Grigonis, E. J. Subak, M. L. Wyrick, Validation Key Practices for Computer Systems Used in Regulated Operations, *Pharmaceutical Technology*, June 1997.

Table 18–1. Period/Events Computer Systems Operational Life.

Period/Event		Operational Life (Period)	
	Early Operational Life	**Maturity**	**Aging**
Representative characteristics		• Phased rollouts • Corrective, adaptive, perfective, and preventive maintenance • Maintainability issues of obsolete technology (e.g., access to electronic records)	
Key Practices	• Problem reporting and maintenance	• Operational audit • Performance evaluation	• Periodic review • Re-engineering analysis

In particular, after a system has been released for use, it should be constantly monitored to uncover any security violations. One must follow up on any security violation, analyze it and take proper action to avoid a reoccurrence.

Another important activity is the process to evaluate new security technologies and their integration with computer systems. The procedural controls implemented as a result of a risk analysis provide a starting point to look for technologies that can replace procedural controls. These procedural controls are the result of security-related implementation requirements identified during the risk analysis.

The security of computer systems is covered in Chapter 15.

Operational Management

Routine use of computer systems requires the following:

- *Training*
  All staff maintaining, operating, and using computer systems that perform regulated operations must have documented evidence of training in their area of expertise. For users, the training will concentrate on the correct use of the computer system, how to report any failure or deviation from normal operating conditions and security. Refer to Chapter 14.

- *Hardware maintenance*
  The procedural controls applicable for the preventive maintenance and repair of the hardware provide a mechanism for anticipating problems and, as a consequence, possible loss of data. Modern hardware usually requires minimum maintenance because electronic circuit boards, for example, are usually easily replaced and cleaning may be limited to dust removal. Diagnostic software is usually available from the supplier to check the performance of the computer system and to isolate defective integrated circuits. Maintenance procedures should be included in the organization's standard operating procedures. The availability of spare parts and access to qualified service personnel are important for the smooth operation of the maintenance program.

Organizations should use replacement parts that meet the design specifications of the original computer system, or the system should be requalified to record that the replacement part performs in accordance with the original specifications.

Computer systems used to control, monitor, or record functions that may be critical to the safety of a product should be checked for accuracy at intervals of sufficient frequency to provide assurance that the system is under control. If part of a computerized system that controls a function critical to the safety of the product is found not to be accurate, then the safety of the product back to the last known date that the equipment was accurate must be determined.

- *Software maintenance*
  Maintenance of a software system includes
  - perfective maintenance or the correction of the system due to new requirements and/or functions
  - adaptive maintenance, or the correction of the system due to a new environment that could include new hardware, new sensors, or controlled devices, new operating systems and/or new regulations
  - corrective maintenance, or the correction of the system due to the detection of errors in the design, logic, or the programming of the system. It is important to recognize that the longer a defect is present in the software before it is detected, the more expensive it is to repair. Refer to Figure 2–2
  - preventive maintenance or the correction of the system to improve its future maintainability or reliability, or to provide a better basis for future enhancements

  As required by the regulations,[2] all maintenance work must be performed after the impact evaluation and approval of the work to be done. Refer to Chapter 13, Change Management.

  It is important to acknowledge that the maintenance of existing software can account for over 70% of all effort expended by a software organization. Refer to Figure 8–2. The computer hardware suppliers normally recommend minimum maintenance schedules including accuracy checks for their components.[3]

- *User support and problem management*
  These enable the reporting and registration of any problem encountered by the users of the system. These can be filtered according to whether their cause lies with the user or with the system itself, and fed back into the appropriate part of the supplier's organization. Those problems that require a possible system change are then managed through a change management procedure.

- *Archiving*
  All SLC documentation should be archived in an environmentally controlled facility that is suitable for the material being archived, and that is both secure and, where possible, protected from environmental hazards. A record of all archived materials should be maintained.

---

[2] For example, European Union (EU) GMP regulations, Annex 11–11.
[3] FDA, *Guide to Inspection of Computerized Systems in the Food Processing Industry*, Office of Regulatory Affairs (ORA), Division of Emergency and Investigational Operations (DEIO) and the Center for Food Safety and Applied Nutrition (CFSAN), http://www.fda.gov/ora/nspect_ref/igs/foodcomp.html.

Business Continuity

Business continuity procedures, including disaster recovery procedures, should ensure minimal disruption in the case of loss of data or any part of the system. It is necessary to ensure that the integrity of the data is not compromised during the return to normal operation. At the lowest level, this may mean the accidental deletion of a single file, in which case a procedure should be in place for restoring the most recently backed-up copy. At the other extreme, a disaster such as a fire could result in loss of the entire system. For this situation a procedure addressing the following should be in place:

- specification of the minimum replacement hardware and software requirements to obtain an operational system and their source
- specification of the period within which the replacement system should be back in production, based on business considerations
- implementation of the replacement system
- steps to revalidate the system to the required standard
- steps to restore the archived data so that processing activities may be resumed as soon as possible

The procedural control employed should be tested regularly and all relevant personnel should be made aware of its existence and trained to use it. A copy of the procedure should be maintained off-site.

Problem Reporting

The procedures to be followed if the system fails or breaks down must be defined and validated. Any failures and remedial actions taken must be recorded. A procedure should be established to record and analyze errors, and enable corrective action to be taken. Refer to the section on computer system incidents located in Appendix E.

Control of Changes

Refer to Chapter 13.

Periodic Review

After being released for use, all computer systems must be periodically reviewed until the system is no longer required for operation. The system documentation, system operation, modifications, deviations, upgrades, and the electronic record management associated with the software application must be reviewed to determine:

- whether the corrective actions from the previous validation/audit were completed
- the level of operational and performance conformance of the system
- whether changes to the system have been properly tested and documented, and whether the cumulative effects of small changes have caused a gradual degradation of system performance and integrity

- the impact of new/revised SOPs and regulations on the current CSV test methodology
- the adequacy of calibration records and the continued suitability of the equipment to which they apply
- whether current operational SOPs are being followed
- an audit of the current security measures that have been established for the system
- the need for any refresher or update training for personnel

The system owner, system administrator, system user, and system support personnel must participate in regular periodic reviews of the validated system. Upon completion of the periodic review, the findings must be documented and an action plan developed to correct any gaps found during the assessment.

A written certification, or other form of verification, must be provided to confirm that the audit and inspection has been implemented, performed and documented, and that any required corrective action has been taken. The certification should be filed in either the system validation or project file.

Retirement

Refer to Appendix E.

Ongoing Verification Program

Computer system I/O errors can occur on validated systems. A computer component (logic circuits, memory, microprocessor) or device (modems, displays), like mechanical parts, can fail after they have been tested. Another source of computer systems I/O malfunctions is electromagnetic interference (radio-frequency interference, electrostatic discharge, and power disturbance). Software errors that were undetected during the development process, may also be the source of I/O errors. In order to detect errors during the operational life of a computer system before it can make decisions using tainted data, an ongoing monitoring program must be established and followed to verify the hardware and software I/Os. As part of the same ongoing monitoring program, stored data should be checked for accessibility, durability, and accuracy.

Chapter 19

Part 11 Remediation Project

INTRODUCTION

As defined in CPG 7153.17, legacy systems are computer systems not in compliance with Part 11. The first fundamental principle of Part 11 is that it requires organizations to store regulated electronic data in its electronic form. The second fundamental principle is the high level of security required to protect regulated electronic records.

As a result of the fundamental principles above, all computer systems performing regulated operations must protect their electronic records by means of access controls, audit trail controls,

> **Regulatory guidance**
>
> *Firms should have a reasonable timetable for promptly modifying any systems not in compliance (including legacy systems) to make them Part 11 compliant, and should be able to demonstrate progress in implementing their timetable. FDA expects that Part 11 requirements for procedural controls will already be in place. FDA recognizes that technology based controls may take longer to install in older systems.*
>
> FDA, CPG 7153.17, *Enforcement Policy: 21 CFR Part 11; Electronic Records; Electronic Signatures*, May 1999.

and Part 11-associated operational checks. For those computer systems that implement electronic signatures, the security and control of electronic signatures must also provided.

As part of the remediation process for legacy systems, a retrospective evaluation should be performed on the electronic records recorded after August 1977 using the legacy systems. The objective is to assess the complete system operation, including: procedural controls, change control, data and system security, any additional development or modification of the software using an SLC approach, the maintenance of data integrity, system backups and operator (user) training.

If evaluation results do not meet the current regulatory requirements, then the retrospective evaluation would not in itself support the trustworthiness of the electronic records. This is an area of concern for electronic records recorded using legacy systems after August 1997.

Figure 19–1 illustrates a complete Part 11 Remediation Project. Like any project, the schedule is based on priorities, time, and the availability of resources.

Figure 19–1. Complete Part 11 Remediation Project.

This chapter identifies the key principles for remediating a noncompliant Part 11 system, but it is not intended to cover everything that an organization's management needs to do in order to achieve and to maintain compliance with Part 11. The PDA/GAMP Forum document[1] is an excellent resource for planning a Part 11 Remediation Project.

EVALUATION OF SYSTEMS

The evaluation of computer systems performing regulated operations is the first phase to achieving an organized, prioritized, and balanced Part 11 Remediation Project approach. The objective of the evaluation is to identify the system's functional and/or procedural gaps; results of the evaluation will determine whether the operational, maintenance, or security procedures specific to the system will provide a controlled environment, which ensures the integrity of the electronic records and/or signatures as stated in the Part 11 requirements.

An evaluation plan is needed in order to define the nature, extent, schedule, and responsibilities of the evaluation and process.

Each system performing a regulated operation must be identified and the operation it performs must be well understood in order to prioritize the work. Data and process flow diagrams may be used as tools for reviewing the operation. Performing Step 2 can obtain the priority rating that is applicable for each system in the Criticality and Complexity Assessment located in Appendix D. Other factors to take into account during the prioritization process are the components and functions that have regulatory implications.

After the computer systems have been prioritized, they are evaluated by following a worksheet similar to that provided in Appendix G. The Part 11 issues to be evaluated are:

1. **Open/Closed systems**
2. **Security**
   a. System security
   b. Electronic signature security
   c. Access codes and password maintenance
      • Access codes and password security
      • Password assignment
   d. Document controls
   e. Authority, operational, and location checks
   f. Records protection
3. **Operational Checks**

4. **Audit Trails**
   a. Audit mechanism
   b. Metadata
   c. Display and reporting
5. **Electronic signatures**
   a. Electronic-signature without biometric/behavioral controls
   b. Electronic-signature with biometric/behavioral controls
   c. Signature manifestation
   d. Signature purpose
   e. Signature binding

An evaluation report must be generated for each computer system, which summarize the current operation of the computer system, allocates its priority, provides a reference to any supporting documentation, and identifies the compliance gaps in the system. Based on the information in the evaluation reports, a Corrective Action Plan can be generated.

---

[1] PDA/GAMP *Good Practice and Compliance for Electronic Records and Signature, Part 2, Complying with 21 CFR Part 11, Electronic Records and Electronic Signatures*, Version 1, September 2001. Published jointly by PDA and ISPE.

CORRECTIVE ACTION PLANNING

The purpose of the Corrective Action Plan is to define the overall activities, schedule, costs, and responsibilities necessary to guide the development and implementation of technological and procedural controls to bring the systems into compliance with Part 11. The plan should identify any existing technological/procedural controls that may be modified or new techno-logical/procedural controls that need to be implemented in order to ensure that the regulatory requirements are completed in a consistent and uniform manner. The remediation action items identified in the Part 11 assessment should be documented in a detailed implementation plan.

Using the evaluation reports, the remediation activities, available resources, and project schedule, the business cost of the remediation approach can be estimated. This will enable a business decision to be made regarding the remediation or replacement of the current system based on the cost-effectiveness of the system and its operational feasibility.

The plan must contain the activities that place emphasis on achieving consistent, high quality, sustainable compliance solutions. The remediation process consists of six major activities. These activities are:

- interpretation
- training
- remediation execution
- new applications assessments
- application upgrade assessments
- supplier qualification program

Once the corrective action plan has been approved, it can then be executed.

REMEDIATION

During the remediation phase, the computer systems are brought into compliance by implementing the procedural and technological controls determined by the Corrective Action Plan. In addition, the processes needed to sustain the compliance solutions are implemented.

Interpretation

The ability to evaluate the risks of a particular system in relation to 21 CFR Part 11 requires a thorough understanding of the regulation and a consistent interpretation. The objective of the interpretation phase of this plan is to provide current, consistent, regulatory interpretation of 21 CFR Part 11 to IT and interested outside parties.

Training

In the context of Part 11, awareness and understanding of the regulation is fundamental to the success of the Part 11 Remediation Plan. The objective of the Part 11 training is to ensure that all system and technical owners have an appropriate level of knowledge of Part 11.

Other forms of training need to be developed in order to support the maintenance of regulatory compliance of computer systems. Refer to Chapter 14 for more information.

Remediation Execution

Once the Corrective Action Plan is approved, the computer technology suppliers and developers are requested to identify how the deficiencies can be overcome. When appropriate, procedural controls need to be developed in order to address the deficiencies that cannot be solved by technological controls.

It is probable that the implementation of technological controls will require a comprehensive SLC including: the recommendation, conceptualization, and implementation of the new technology; the release and early operation of the new technology; and the decommissioning and disposal of the old technology. If the technological implementation fails, the failure should be documented along with details of the corrective action taken. Once this action has been taken, the system must be re-evaluated in the same way as any other system that had been subject to an upgrade or correction.

When all of the action items applicable to a computer system have been implemented, it can be formally released for operation, and for support under a maintenance agreement. Refer to Chapter 18.

The Corrective Action Plan should be periodically reviewed because the evolving technology requirements will need to be considered and the plan revised accordingly.

New Applications and Application Upgrade Assessments

The objective of Part 11 assessments for new applications, and for application upgrades, is to identify the Part 11 gaps before releasing the system into production. All gaps must be managed using procedural and/ or technological controls.

New systems and upgrades to systems that are released into production must have a high level of compliance. However, due to limitations on the technology used by suppliers, it is not always possible to implement systems that are fully compliant.

These systems must be assessed, the Part 11 gaps recorded, and a formal plan established to remediate the gaps.

The project team is responsible for completing this assessment.

Suppliers Qualification Program

A key business strategy has been the outsourcing of work to computer technology suppliers and developers. The objective of qualifying computer technology suppliers and developers is to evaluate and monitor these 'strategic' partners for Part 11 compliance and to provide an input to the partner selection and partner relationship management processes.

For each qualification performed, a report must be prepared that describes the results of the qualification.

REMEDIATION PROJECT REPORT

The Remediation Project Report provides evidence of successful project completion. It must describe the technological and procedural controls and associated activities necessary to make the computer technologies that were assessed compliant with Part 11. This report and all supporting documentation should be archived.

Once a legacy system has achieved a satisfactory, documented Part 11 compliant state, any subsequent changes can be prospectively validated.

Chapter 20

Operational Checks

The objective of operational checks is to enforce the sequencing of steps and events as appropriate. The application-dependent algorithms consisting of sequencing of operations, instructions to the operator, critical embedded requirements, and safety related precautions, are encompassed in the computer program(s) driving the computer systems. These application-dependent algorithms and Part 11 requirements become part of the computer systems validation model (refer to

Regulatory guidance

Use of operational system checks to enforce permitted sequencing of steps and events, as appropriate.

Department of Health and Human Services, Food and Administration, *21 CFR Part 11, Electronic Records; Electronic Signatures*, Federal Register 62 (54), 13430–13466, March 20, 1997.

Chapter 4 for more information) through the 'operation' process input refer to Figure 4–1.

The lack of operational checks to enforce event sequencing is significant if an operator deviates from the prescribed order of computer system operation steps, resulting in an adulterated or misbranded product, and/or data integrity.

There are three types of operational checks: instructions to operators, system sequencing, and Part 11 operational checks.

INSTRUCTIONS TO OPERATORS

At the operator level, the purpose of operational checks is to guide the operator during specific actions. The intention is to prevent operators from operating the system and/or signing records outside of the preestablished order. The preestablished order is specified in the system requirements specification. Two examples of typical operator actions that require the enforcement of a sequenced approach include:

- Signing a batch record to show that the addition of an ingredient occurred before the ingredient was actually added
- Verification by a second person that an activity was performed as specified by the batch record.

OPERATION SEQUENCING

Operational checks are normally presenting process control computer systems. These systems may contain code that is part of the master production record. At the system level, the purpose of operational checks is to execute algorithms, sequencing of operations, and safety-related functions as required in the applicable customer specification. Inspections and testing are fundamental processes to be performed during the validation of critical system sequences. In addition, an ongoing program must be established to frequently verify that critical operations occur in the proper sequence.

Section 211.188(b)(11) of the Current Good Manufacturing Practice Regulations is an example of operation sequencing. This section requires that batch production and control records include documentation that each significant step in the manufacture, processing, packing, or holding of a batch was accomplished, including identification of the persons performing, directly supervising or checking each significant step in the operation.

When the significant steps in the manufacturing, processing, packing, or holding of a batch are performed, supervised, or checked by a computerized system, an acceptable means of complying with the identification requirements of 21 CFR 211.188(b)(11) would consist of conformance to all of the following[1]:

- documentation that the computer program controlling step execution contains adequate checks, and documentation of the performance of the program itself
- validation of the performance of the computer program controlling the execution of the steps
- recording specific checks in batch production and control records of the initial step, any branching steps and the final step

In general, assessment to the compliance to 211.188(b)(11) includes documented demonstration that the computer system examines the same conditions that a human being would look for, and that the degree of accuracy in the examination is at least equivalent.

Other examples of typical system sequencing include the correct automatic operation of equipment controlled by a computer system such as fluid bed system, automated packing lines, and product release is another function that involves sequencing.

PART 11-RELATED OPERATIONAL CHECKS

Part 11 contains specific sequencing requirements that must be implemented using computer technologies. These operational checks usually fall in the category of operation sequencing and are built into the software/hardware. To highlight the importance of these operational checks, the author suggests that specific categories are allocated for them as follows:

- *Electronic signature manifestation, 11.50(b)*
  The electronic signature must be displayed in any human readable form, including printouts and video displays:
  - immediately after the signature is executed
  - after displaying a signed record
  - when printing signed electronic record(s)
  The printed name of the signer, the date and time of signing, and the meaning associated with the signing must be displayed.

- *Multisigning, 11.200(a)(1)(i)*
  When an individual executes one or more signings that are not performed during a single, continuous period of controlled system access, each signing must be executed according to the following:

---

[1] CPG 7132a.08, *Identification of 'Persons' on Batch Production and Controls Records*, November 1982.

- 1st signing: using both the userID and the password components
- 2nd + and subsequent signings during a period of continuous, controlled access: either re-entry of the password alone, or use of both the userID and password components
- One signature can apply to multiple data entries on a screen as long the items that the signature applies to be clearly indicated

- *Flagging unauthorized use of passwords and userID codes, 11.300*
  Systems that use electronic signatures must be designed so that unauthorized attempts to use the signature are detected and reported to security management. The security system should be capable of identifying situations where misuse occurs and should notify security management appropriately. Repeated attempts at unauthorized access must result in access to the software application being automatically disabled.

  A notification option may be used to log the message to a historical file or to send a message to a system administrator workstation. A process for investigating attempted security violations must be defined so that security violations are handled promptly.

- *Authority checks*
  The system must implement authority checks to ensure that only authorized individuals can use the system to sign records, access the operation or device, alter records, or perform the operation at hand, 11.10(g). These authority checks are based upon the various roles and responsibilities assigned to individuals who are known to the system.

  The computer system must be designed to make distinctions between the control access to the system, to functions in the system, and to input and output devices used by the system.

- The system shall be designed to implement device checks (11.10(h)) including recording the location (node) of the workstation where each entry was made. Device checks enable the software application to determine whether the input being generated by a particular device is appropriate. The use of the term 'as appropriate' does not require that device checks are performed in all cases. These checks may be used when certain devices have been selected to be legitimate sources of data input or commands. For example, in a network environment it may be necessary for security reasons to limit the issuance of critical commands to a particular authorized workstation.

- The application must detect when a workstation has experiences a long idle period and must be able to automatically log out the user.

- Audit trails refer to a journal or record of modifications to electronic records by the users or by processes operating on the users behalf for tracking purpose. Data must be protected from unauthorized modification and destruction to enable the detection, and after-the-fact investigations, of security violations.

  This mechanism provides the capability for modified data to be reconstructed in its previous form and therefore the modification must not obscure the previously entered data. The tracking mechanism must include a computer-generated time stamp that indicates the time of the entry that modified the record and the types of modifications performed. The requirements for audit trails are as follows:
  - they must be computer generated
  - they can be either part of the electronic record itself or a separate record
  - they must not be modified by the 'person' who create them
  - they must indicate, 11.10(c):

> ➢ when the data was first entered and by whom
> ➢ when who made to the data and any changes

- *Signature/record linkage must be computer generated, 11.70*
  The linkage may be achieved by linking the userID, which could be obtained from a password/security file. The signatures must not be capable of being removed, copied, changed, or transferred in order to falsify an electronic record by ordinary means. The signer's full name doesn't have to be embedded in the record itself. The field containing the name can be a pointer that points to a file containing the full names of the signer.

  The link must be retained for as long as the record is kept, just as a handwritten signature stays with the paper, long after an employee has departed a company. Although a userid/password can be removed from a current user database, it must still be retained in an archive in order to maintain the signature and record linkage.

VALIDATION OF OPERATIONAL CHECKS[2]

An acceptable method for the validation operational checks would be as follows:

1. The documentation of the program, including a requirements specification that describes what the software is intended to do
2. The performance of inspections and testing so that no step or specification can be missed or poorly executed/assigned
3. Documentation of the initial and final steps

---

[2] FDA, *Guide to Inspection of Computerized Systems in Drug Processing*, February 1983.

Chapter 21

Compliance Policy Guide (CPG) 7153.17

INTRODUCTION

The main objective of Compliance Policy Guide (CPG) 7153.17 is to provide specific guidance to the FDA field investigators on whether or not to pursue regulatory action against an organization under Part 11. It establishes the direction to be followed by these investigators when they find systems that are not properly compliant with Part 11. CPG 7153.17 makes the following points:

Regulatory guidance
FDA will consider regulatory action with respect to Part 11 when the electronic records or electronic signatures are unacceptable substitutes for paper records or handwritten signatures, and that therefore, requirements of the applicable regulations (e.g. CGMP and GLP regulations) are not met. Regulatory citations should reference such predicate regulations in addition to Part 11. FDA, CPG 7153.17, *Enforcement Policy: 21 CFR Part 11; Electronic Records; Electronic Signatures,* 5/13/99.

- The FDA defines 'legacy systems' as those systems in place before August 20, 1997.[1]
- The main area of enforcement of this CPG comes from the 'no grandfathering' clause. This clause applies to legacy systems and the electronic records stored by these systems. The 'no grandfathering' clause indicates that Part 11 provisions relating to record creation do not cover electronic records that are created before the effective date of the rule. These records would not need to be retrospectively altered.
- The four elements to be considered by the field inspectors when pursuing regulatory actions are
  - nature/extent of the deviation
  - impact on product quality/data integrity
  - adequacy/timeliness of the corrective action plan
  - compliance history

Nature/Extent of the Deviation

FDA inspections are conducted for many reasons and the extent of compliance of computer systems with Part 11 may be evaluated during any of these inspections. If applicable, a field investigator will review electronic records in order to evaluate their level of compliance with the predicate rule. At the same time, the investigator may check adequacy of record keeping arrangements for compliance with Part 11.

Some of the elements to be evaluated are:

---

[1] FDA, Human Drug CGMP Notes, Vol. 7, No. 3, September 1999.

- Does the system meet the appropriate Part 11 requirements?
- Has the system been validated?
- Have Part 11 procedural controls been met?
- Has the company submitted certification for electronic signatures?

Audit trails are the most important technical consideration and the most important element of Part 11. These are generated when data, which is stored on durable media, is modified. Audit trails ensure that changes to the electronic records do not obscure the source data. Electronic data as well as data collected manually can be amended and the amendment must be recorded and justified[2]. Changes to electronic records will always require an audit trial[3]; these changes must not obscure the original record. Audit trails must include the original entry, the new entry, the reason for the change, the date and the time of the change, and the electronic signature.

Audit trails may be a special consideration during a recorkeeping inspection. The requirements for audit trails are as follows:

- They must contain who did what, wrote what, and when (date and time stamp)
- They can be part of the electronic record or a record by itself
- They must be computer generated and saved electronically in chronological order
- They must be created when the operator performs entries and actions that create, modify, or delete electronic records, e.g., command to acknowledge an alarm
- They must not be modified by the 'person' who creates them

Impact on Product Quality/Data Integrity

This element of guidance to be used by the FDA inspectors is based on considerations such as the GxP criticality of the computer system.

The 'criticality,' or product risk, must be assessed for both the direct and indirect impact of the computer system on the product. The extent of the validation/qualification activities to be carried out will depend on the results of this assessment. The criticality is based on the potential impact on the product, and existing data does not justify a lower rating.

Other factors that should be considered to assess product impact are:

- manufacturing
- labeling
- packaging
- master batch records/device history records
- quality records (the data collected and used to make product release decisions)
- product performance (product and manufacturing)
- delivery
- inventory

The electronic data obtained from the areas mentioned above are highly critical. The field inspector will look for evidence of any lack of control associated with the electronic records.

---

[2] Final Rule for 21 CFR Part 211.100 (1996), *Current Good Manufacturing Practice For Finished Pharmaceuticals*.
[3] 21 CFR 11.10(e).

Adequacy/Timeliness of the Corrective Action Plan

The corrective action plan may be contained within a validation plan. The validation plan is a document that describes the company's overall philosophy, intentions, timetable, and approach to be followed for the corrective action compliance program. The approach to be taken may consist of writing procedures, performing verifications, and/or performing qualification activities. Refer to Chapter 19 for information on remediation project.

The FDA expects procedural controls relating to Part 11 to be in place. These include Change Control, Security, System Start-Up/Shutdown, Data Collection and Management, Backups, System Maintenance, System Recovery, Training, Periodic Reviews, Revalidation, Contingency Planning, Performance Monitoring, and Decommissioning.

These procedural controls must be verified during the qualification of the computer system and periodically evaluated.

Compliance History

The compliance history of the company being inspected, especially with regard to data integrity, will be taken into account in order to 'intensify their scrutiny of electronic records.' If a company has a history of Part 11 violations or inadequate or unreliable record keeping, this would make any additional Part 11 deviations more significant. Until firms reach full compliance, FDA investigators will be more vigilant when attempting to detect inconsistencies, and unauthorized changes.

Chapter 22

Electronic Records

REGULATORY GUIDANCE[1]

Electronic record means any combination of text, graphics, data, audio, pictorial, or other information representation in digital form that is created, modified, maintained, archived, retrieved, or distributed by a computer system. 21 CFR Part 11.3(6).

WHAT CONSTITUTES AN ELECTRONIC RECORD?

An electronic record is any information that is recorded in a form that:

- requires a computer or other machine to process the electronic record[2] and that satisfies the legal definition of a record in 44 U.S.C. 3301[3]
- only a computer can process[4] and that satisfies the definition of a record in 44 U.S.C. 3301

The definitions above are consistent with Part 11. Per Part 11, paragraphs 22, 45, and 72 in the preamble refer to when data on transient memory become electronic records. In summary, the primary attribute to take into consideration for the above definitions is record 'retrievability.' It is essential to take this into account when deciding the applicability of Part 11.

The current federal legislation, Electronic Signatures in Global and National Commerce Act, defines electronic records in more general terms than Part 11.

---

[1] FDA, *Electronic Records; Electronic Signatures Final Rule*, 62 Federal Register 13430 (March 20, 1997).

[2] US Department of Defense, *DOD 5015.2-STD — Design Criteria for Electronic Records Management Software Applications*, November 1997.

[3] 'Records' includes all books, papers, maps, photographs, machine readable materials, or other documentary materials, regardless of physical form or characteristics, made or received by an agency of the United States Government under Federal law or in connection with the transaction of public business and preserved or appropriate for preservation by that agency or its legitimate successor as evidence of the organization, functions, policies, decisions, procedures, operations, or other activities of the Government or because of the informational value of data in them. Library and museum material made or acquired and preserved solely for reference or exhibition purposes, extra copies of documents preserved only for convenience of reference, and stocks of publications and of processed documents are not included.

[4] 36 CFR 1234.2.

For this definition, the primary attribute to take into consideration is whether the data is 'accessible' once it is put into storage, not the 'retrievability' of the record. This definition would consider data in transient memory to be an electronic record.

WHAT CONSTITUTES A PART 11 REQUIRED RECORD?

Part 11 required records are those:

- that are defined in the applicable predicate rule, especially those used to determine product quality
- that may not be explicitly identified in the applicable predicate rule, but is a requirement that the organization imposes upon itself

One example of this 'self-imposition' by an organization might be the records required by procedural controls such as those for master batch records.

HOW SHOULD PART 11 RECORDS BE MANAGED?

Required electronic records that meet Part 11 are used to satisfy the applicable predicate rule. The following are some basic principles that must be adhered to:

- Procedural controls must be established and maintained providing instructions for identification, collection, indexing, filing, storage, retention requirements, maintenance, and disposition of quality system records. (Section 5.1, FDA, draft of *Guidance for Industry: 21 CFR Part 11; Electronic Records; Electronic Signatures: Maintenance of Electronic Records*, July 2002)
- Computer environments that contain and manage such records must be validated. (Part 11 and EU GMPs Annex 11–2)
- Detailed procedures relating to the system in use should be available and the accuracy of the records should be checked. (EU PIC/S and Section 5.3, FDA, draft of *Guidance for Industry: 21 CFR Part 11; Electronic Records; Electronic Signatures: Maintenance of Electronic Records*, July 2002)
- After the computer environment has been validated, there should be an ongoing monitoring program to verify the data/control/ monitoring interface(s) between the system and external equipment in order to ensure the correct input and output transmission of data (CPG 7132a.07)
- As part of an ongoing monitoring program, stored data should be checked for accessibility, durability, and accuracy (EU GMPs Annex 11–13)
- Electronic records should be secured by physical or electronic means against wilful or accidental damage. Electronic records should only be entered or amended by persons authorized to do so (EU GMPs Annex 11–13, 8)
- Data should be protected by backing-up at regular intervals (EU GMPs Annex 11–14)
- There should be available adequate alternative arrangements for systems that need to be operated in case of a breakdown. The procedures to be followed if the system fails or breaks down should be defined and validated (EU GMPs Annex 11–5, 16)
- Any alteration to a required record should be recorded with the reason for the change (EU GMPs Annex 11–10)

- There should be a defined procedure for the issue, cancellation, and alteration of authorization to enter and amend data, including the changing of personal passwords (EU GMPs Annex 11–8)
- If changes are proposed to the computer equipment or its programs, the checks mentioned above should be performed at a frequency appropriate to the storage medium being used
- It is particularly important that the electronic records are readily available throughout the period of retention (EU PIC/S and Section 5.2, FDA, draft of *Guidance for Industry: 21 CFR Part 11; Electronic Records; Electronic Signatures: Maintenance of Electronic Records*, July 2002)

MINIMUM RECORD RETENTION REQUIREMENTS

Part 11 defines how electronic records must be managed. The retention requirements of such records are contained in the applicable predicate regulation. Table 22–1 refers to some of the sections in the predicate regulations or guidelines that specify record retention requirements.

Table 22–1. Records Retention Requirements.

Area	Section	Regulation
GLP	Organization and personnel records	FDA 58.3
GLP	Specimen and data storage facilities	FDA 58.51
GLP	Storage and retrieval of records and data	FDA 58.190 & 58.195
GLP	Organization and personnel	OECD 1.1 & 1.2
GLP	Sampling and storage	OECD 6.1
GLP	Archive facilities	OECD 3.4
GLP	Characterization	OECD 6.2
GLP	SOPs	OECD 7.2
GLP	Storage and retrieval	OECD 10.1 & 10.2
GLP	Storage and retrieval with archive facilities	OECD 10.2
GCP	Supply and handling of investigational product(s)	ICH 5.14
GCP	Trial management, data handling, record keeping	ICH 5.5
GMP	Records and documentation	FDA 211.180
GMP	Personnel records	EU PIC 2
GMP	Documentation	EU PIC 4

The organization owning the electronic records may impose more stringent retention requirements that may be based on legal requirements.

WHEN ARE AUDIT TRAILS APPLICABLE FOR ELECTRONIC RECORDS?

The point at which data becomes an electronic record should reflect its origin and intended purpose. For manufacturing and laboratory records this may be the initial point of data acquisition but for a clinical trail report it may be from the report's first official use. Electronic records can be divided into three categories based on their intended purposes: instructions, events, and reviews. Using these categories, it is possible to determine when audit trails are applicable. Table 22–2 summarizes these concerns and needs.

INSTRUCTIONS

The instructions embedded in computer program(s) either provide information on what the operators are supposed to do, or information on how the equipment or processes are intended to function. For example, for manufacturing operations, the batch production and control records may provide procedures, controls, instructions, specifications, and precautions to be followed when using computer systems. These programs may also contain control data for product formulation, batch size, yields, and automated in-process sampling/testing procedures.

Instructions may be modified before being loaded into the associated workstation. The modification may be related to the adaptation of a process or the introduction of a new set of instruction to the operator. In either case, the modification must be qualified and the operators trained to use it.

Instructions are version controlled and need to go through a formal approval process before they can be used. When working with instructions in electronic format, the audit trails should start when the instruction is official approved.

In a 'paper-based environment,' if a procedural control is, in fact, a draft, (e.g., an early version of a procedural control that is ultimately not adopted) the applicable predicate rule would probably not require that it be retained. The same approach is also applicable to instructions in electronic format.

When there is a deviation from an instruction, the system must have the capability to record the deviation and the associated audit trail must contain the action(s) carried out by the operator. In addition, an investigation should be carried out and the cause of the deviation justified.

EVENTS

Events are the various defined actions or activities performed by personnel, equipment, or instruments. The records of events provide evidence that make it possible to establish what

Table 22–2. E-records Concerns and Needs.

Record Type	Key Concern	Key Control
Instruction records (records that establish or define expectations, e.g., Validation Plans, Protocols, SOPs, specifications, recipe, method)	• Is this the latest approved version? • What version was used for that specific batch or experiment?	• version and distribution management • control of change • access security
Event records (the information is recorded contemporaneously with the actions, e.g., electronic data from laboratory instruments or manufacturing control operations)	• Who? When? What? • Reason of changes or corrections?	• audit trail
Review records (records that provide evidence that events, information, or data have been evaluated for content, completeness, or accuracy, e.g., reviews required in 211.192. Usually these records are approved.)	• Who takes responsibility? • Responsibility for what? • Review, approval, ...? • Based on what data?	• signature requirements • record/signature link

actually happened during a process. These types of documentation cover important aspects that need to be recorded in order to demonstrate that systems performing regulated operations are adequately controlled. For example, for a computerized drug process, certain information required by the CGMP regulations to be included in a master production record is contained within the source code for the application program.

For process equipment controlled by computers, the events are recorded at the time of occurrence by the computer systems. Instrumentation connected to computer systems may measure process parameters or document actions such as addition of ingredients as a specific process event. The recording of processing events and operator actions must be audit trailed immediately, and should not wait for approval.

As a consequence of the distinctiveness of electronic records resulting from events, one area of awareness is amendment and (logical) deletion of these electronic records. Procedural controls must define the process of correction. The person supervising or checking the operation should perform amendments to event files.

REVIEWS

Reviews are written evidence that events, instructions, information, or data have been evaluated for content, completeness, or accuracy. Critical reviews are those where acceptance criteria are compared against the results of a process to ensure that the data conforms to requirements.

An example of a required review is the one mandated by Section 211.188(b)(11) of the CGMP regulations. It is a requirement that batch production and control records include documentation that verified that each significant step in the manufacture, processing, packing, or holding of a batch was accomplished, including the identification of the persons performing, directly supervising, or reviewing each significant step in the operation. The intention of the regulation is to ensure that each significant event in a process is reviewed and that there is record evidence to show this. It is quite possible that a computer system could achieve either the same or higher degree of review assurance. In this case, it may not be necessary to specifically record the checks made on each series of steps in the operation.

Reports may be considered to be reviews. For electronic documents that provide an interpretation of data drawn from electronic records that are already subject to control, there is no clear need to reimpose these controls until the document is used for some official purpose. In a case of submissions, once a submission has been assembled and approved ready for submission to a regulatory agency it would then become an electronic record subject to audit trails.

If any electronic record resulting from an event, instructions, information, or data that was part of a review is modified, the approver and reviewer must be prompted to repeat the review of the data. This repeat review must be prompted by procedural or technological controls.

PRESERVATION STRATEGIES[5]

While electronic records enable information to be created, manipulated, disseminated, located, and stored with increasing ease, preserving access to this information poses a significant

---

[5] D. Bearman. *Functional Requirements for Recordkeeping, in Electronic Evidence: Strategies for Managing Records in Contemporary Organizations*, Pittsburgh: Archives & Museum Informatics, 1994, http://www-prod.nla.gov.au/padi/topics/thesaurus.html.

challenge. Unless preservation strategies are actively employed, this information will rapidly become inaccessible. Choice of strategy will depend upon the nature of the material and what aspects are to be retained.

Copying information without changing it, offers a short-term solution for preserving access to digital material by ensuring that information is stored on newer media before the old media deteriorates beyond the point where the information can be retrieved.

The migration of digital information from one hardware/software configuration to another or from one generation of computer technology to a later one, offers one method of dealing with technological obsolescence.

While adherence to standards will assist in preserving access to digital information, it must be recognized that technological standards themselves are evolving rapidly.

Technology emulation potentially offers substantial benefits in preserving the functionality and integrity of digital objects. However, its practical benefits for this application have not yet been well demonstrated.

Encapsulation, a technique for grouping together a digital object and anything else necessary to provide access to that object, has been proposed by a number of researchers as a useful strategy in conjunction with other digital preservation methods.

The importance of documentation as a tool to assist in preserving digital material is universally agreed upon. In addition to the metadata necessary for resource discovery, other sorts of metadata, including preservation metadata, describing the software, hardware, and management requirements of the digital material, will provide essential information for preservation.

The requirement for keeping every version of software and hardware, operating systems, and manuals, as well as the retention of personnel with the relevant technology skills, generally makes the preservation of obsolete technologies unfeasible.

Various frameworks designed to assist in managing the preservation of digital material have been developed. These include tools designed to assist in the development of digital preservation strategies. These will often entail the identification of the various stages in the operation and future planning of a computer system when the provision of long-term access should be considered.

ELECTRONIC RECORDS AUTHENTICITY[5]

The authenticity of electronic records refers to the degree of confidence that users can have that the records are the same as those expected, based on a prior reference or understanding of what they purport to be.

The digital environment poses particular challenges for establishing authenticity. This is due to the ease with which electronic records may be altered and copied, which can result in the possibility of a multiplicity of versions of a particular document.

Methods used in converting, storing, transmitting, or rendering electronic records may result in distortions and therefore need to be documented. The process of migrating information from one system or format to another may result in changes, which also need to be recorded.

Aspects such as a document's functionality, its dependence on a particular software application, and its relationship to other documents are all features that need to be considered in the establishment of its authenticity.

A range of strategies for asserting the authenticity of digital resources have been developed, and the choice of a particular method will depend upon the purpose for which authenticity verification is required. Among these methods are the registration of unique document identifiers and the inclusion of metadata within well-defined metadata structures. Hashing and digital time stamping are 'public' methods, authenticating the existence of a document relating to a specific moment in time. Another class of methods for establishing authenticity includes encapsulation techniques and encryption strategies. A digital watermark can only be detected by appropriate software, and is primarily used for protection against unauthorized copying. Digital signatures arc used to record authorship and identify the people who have played a role in a document.

STORAGE[5]

The choice of strategies and methods for storing digital information will determine its future accessibility and usability. A comprehensive enterprise-wide data warehouse is the subject-oriented, integrated, time-variant, nonvolatile collection of data, which can serve as the central point of data integration for business intelligence and supports strategic, tactical, and operational decision-making processes across the entire enterprise.

Options for storing digital information include the choice of media on which the information will be stored, and decisions regarding the formats in which it will be stored.

The physical medium on which the information is stored should, ideally, be stable: deterioration can be slowed by storing the digital information under stable environmental conditions. General information regarding the preservation of access to information on physical format digital materials is available on the National Library of Australia's Preserving Access to Digital Information (PADI) site (http://www-prod.nla.gov.au/padi/topics/thesaurus.html), as well as along with more specific information about preserving access to material stored on optical disks and magnetic media.

Strategies such as storing identical material in multiple locations and regularly backing up will provide some protection against loss due to media failure, physical damage to facility/property, or human error.

While the choice of standard storage media can ameliorate the effects of technological obsolescence, inevitably, old storage media will be superseded and migration, or some other preservation strategy will be necessary to preserve access to the material. Similarly, the selection of less volatile standard formats will assist in maintaining access and will facilitate later migration; however, it must be remembered that in the information technology environment, the standards themselves will change.

Whether digital information is stored on-line, near-line, or off-line will usually depend upon the expected retrieval requirements. Little used off-line material may be stored on magnetic tape, while the enhanced accessibility of disk storage may be chosen for high-demand on-line material.

Digital information stored in compressed form requires less space and thus reduces storage costs. However, when an image is compressed and then decompressed, the decompressed image is usually not quite the same as the original scanned image; this is called 'lossy compression.' Distortions can be particularly severe for high-compression ratios; although the degree of loss can usually be reduced by adjusting the compression parameters. The use of 'lossy compression' for the long-term storage of digital material is not recommended, due to the potential for irretrievable loss of information on decompression, or on the migration of the digital material from one 'lossy compression' system to another.

Chapter 23

Electronic Signatures

REGULATORY GUIDELINE[1]

Electronic signature means a computer data compilation of any symbol or series of symbols executed, adopted, or authorized by an individual to be the legally binding equivalent of the individual's handwritten signature. 21 CFR Part 11.3(7)

GENERAL CONCEPTS

The application of an electronic signatures refers to the act of affixing, by electronic means, a signature to an electronic record. Part 11 references two types of nonbiometrics-based electronic signatures: password/userid combination-based signatures and digital signatures. In addition to these methods, authentication systems (e.g., passwords, biometrics, physical feature authentication, behavioral actions, and token-based authentication) can be combined with cryptographic techniques to form an electronic signature. Refer to Chapter 15 for additional information on authentication.

Whichever method is used, there are some basic elements to be considered.

1. An electronic signature solution must make electronic signatures secure through the use of a copy protection mechanism that makes it impossible to copy, cut, or paste signatures and audit trails from an approved record. This is an element that is necessary in order to ensure the integrity of digitally signed records.
2. In the electronic environment, an electronic signature on an electronic record must carry the same legal weight as an original signature on a paper-based document.
3. The electronic signature process involves:
   * authentication of the signer
   * a signature process that complies with the system design and software instructions specified
   * the binding of the electronic signature to the electronic record
   * nonalterability after the signature has been affixed to the electronic record
4. The controls applicable to electronic signatures include:
   * uniqueness of the signature

---

[1]   FDA, *Electronic Records; Electronic Signatures Final Rule*, 62 Federal Register 13430 (March 20, 1997).

- signature/record linking
- electronic signature security
- password management (e.g., assignment, removal, loss management, aging)
- in the case of dynamic passwords, it is required that the tokens that generate the password are periodically tested

5. Electronic signature manifestation

The electronic signature must be displayed in human readable form, including printouts and video displays:

- immediately after the signature is executed
- after displaying a signed record
- when printing signed electronic record

The printed name of the signer, the date and time of the signing, and the meaning associated with the signing must be displayed.

6. Multisigning

When an individual executes one or more signings that are not performed during a single, continuous period of controlled system access, each signing must be executed according to the following:

- first signing: using both the userID and the password components
- second and subsequent signings during a period of a continuous, controlled access: either re-entry of the password alone or using both the userid and the password components
- One signature can apply to multiple data entries on a screen as long as the items that the signature applies to are clearly indicated

PASSWORD-BASED SIGNATURES

In combination with PINs, Part 11.300 allows the use of password-based signatures. As stated in Chapter 15, there are two PIN/password based authentication schemes: static passwords and dynamic passwords. The same PIN/password combination used for authentication may also be used for an electronic signature. The affixing of a signature to a record should be an affirmative act that is deliberate, unique, and independent of the authentication process, and that serves the ceremonial and approval functions of a signature and establishes the sense of having legally consummated the transaction.

The records/signatures linking using password/PIN-based signatures is either centered on the use of software locks, the storage of the electronic signature in a database table separate from its associated record, or the storage of the signature within the subject electronic record.

DIGITAL SIGNATURES[2]

Digital signatures are a form of electronic signatures suggested for open systems. A digital signature provides the mechanism for verifying the integrity of the signature/record linkage, and the identity of the signatory. The signature/record linkage is a fundamental requirement for conformance with 21 CFR Part 11.70. Digital signatures can be implemented in software, firmware, hardware, or any combination of these items. Table 23–1 depicts the key features of

---

[2] O. López, *Overview of Technological Controls Supporting Security Requirements in Part 11,* presented at the PDA Annual Meeting, Washington, D.C., December 2001.

Table 23–1. Electronic Signature.

Requirement	Implementation
Digital signature If a digital signature is employed, three features must be implemented: Message integrity, nonrepudiation, and user authentication. Other implementation features are optional.	• Ability to add attributes • Continuity of signature capability • Counter signatures • Independent verifiability • Interoperability • Message integrity • Multiple signatures • Nonrepudiation • Transportability • User authentication

digital signatures. Various technologies may fulfill one or more of the requirements specified in Table 23–1. A comprehensive implementation requires the assurance of the following:

- records integrity
- nonrepudiation
- user authentication

The PKCS[3] describe how to digitally sign a message/record in such way that the recipient can verify who signed it, and verify that it has not been modified since it was signed. Refer to Figure 23–2. In summary:

1. The sender's digital signature is associated with a pair of keys:
   - **Private key**
   - **Public key**
2. In order to sign a record, the record and the private key form the inputs to a hashing process.
3. The output from the hashing process is a string bit pattern (**message digest**) that is appended to the record. The plain text, the digital signature, and the sender's digital signing certificates[4] are sent to the recipient.
4. At the recipient site, after the sender's certificate is received, the CA digital signature is checked to ensure that it was issued by someone that the recipient trusts.
5. The recipient of the transmitted record decrypts the **message digest** with the originator's public key, applies the same message hash function to the record, and then compares the resulting **message digest** with the transmitted version.
6. Any modification to the record after it was signed will cause the signature verification to fail (*integrity*).
7. If the digital signature was computed with a private key other than the one corresponding to the public key used for verification, then the verification will fail (*authentication*).

---

[3] Public-Key Cryptography Standards (PKCS) are a family of standards for public-key encryption developed by RSA Laboratories. It describes the syntax for a number of data structures used with public-key cryptography.

[4] A signing certificate contains the public signing key assigned to an individual.

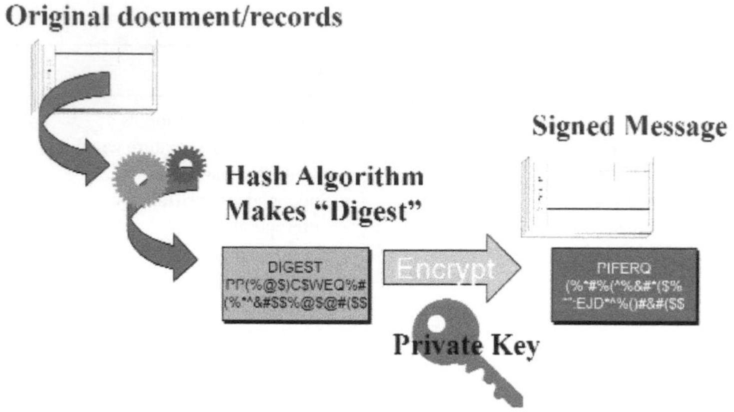

Original document/records

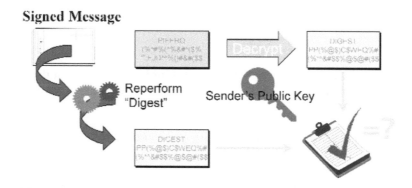

Figure 23–1. Digital Signature Process.

Figure 23–2. Signature Verification Process.

In digital signatures, the private key signs, and the public key verify the authenticity of signatures. For confidentiality, the public key encrypts messages, and the private key decrypts messages.

Digital signatures are technologies that fully support the trustworthiness of signed records. Some products on the market that support the integrity and signature authentication to documents written in Microsoft Word®, Microsoft Excel®, and Microsoft Outlook®, Adobe Acrobat®, JetForm FormFlow, PureEdge, XML, and HTML.

Some digital signature standards are:

- Rivest-Shamir-Adleman (RSA) as specified in ANSI X9.31 Part either 1 (ISO 9796) or PKCS #1
- Digital Signature Standard (DSS) as specified in ANSI X9.30 Part 1 and NIST FIPS PUB 186-2
- Elliptic Curve DSA (ECDSA) as specified in ANSI X9.62

Chapter 24

Technologies Supporting Part 11[1,2]

PAPER-BASED VERSUS ELECTRONIC-BASED SOLUTIONS

The record requirements discussed in Chapter 22 apply to both hard-copy records and to electronic records. The privacy, authenticity, reliability, and nonrepudiation mechanism used in a traditional paper-based solution consist of:

Condition	Solution
Privacy	Envelopes
Authenticity	Notaries, strong ID, physical presence
Reliability	Signatures, watermarks, barcodes
Nonrepudiation	Signatures, receipts, confirmations

The privacy, authenticity, reliability, and nonrepudiation mechanisms used in a digital-based solution consist of:

Condition	Solution
Privacy	Data encryption
Authenticity	Digital signatures, digital certificates
Reliability	Hash algorithms, message digests, digital signatures
Nonrepudiation	Digital signatures, audit trails

Electronic-based solutions can be used to support the security requirements contained in Part 11. The following sections in this chapter provide an overview of the main electronic-based solutions necessary to achieve trustworthy records, and to securing computer resources to manage these records.

As will be noted, these solutions do not support the technological controls necessary to implement audit trails.

---

[1] O. López, Technologies Supporting Security Requirements in 21 CFR Part 11, Part I, *Pharmaceutical Technology*, February 2002.
[2] O. López, Technologies Supporting Security Requirements in 21 CFR Part 11, Part II, *Pharmaceutical Technology*, March 2002.

HASH ALGORITHMS

Hashing refers to the process of computing a condensed message or record of any length into a fixed length string using a one-way mathematical function so that one cannot retrieve the message from the hash. The output of hashing process is called a message digest. The probability of two different records generating the same message digest is 1 in 1087. Consequently, a message digest is unique and has a low probability of collisions. Any change in a message, however minor, will result in a change in the message digest.

Because hashing is a one-way function and the output of the function has a low probability of collisions, hashing can be used with the cryptographic product or services family for authentication, nonrepudiation, and data integrity. An example of these is the Digital Notary.[3] Hashing is also a key element in the DSA.

> **Regulatory guidance**
>
> *Persons who use open systems to create, modify, maintain, or transmit electronic records shall employ pro-cedures and controls designed to ensure the authenticity, integrity, and, as appropriate, the confidentiality of electronic records from the point of their creation to the point of their receipt. Such procedures and controls shall include those identified in Sec. 11.10, as appropriate, and additional measures such as document encryption and use of appropriate digital signature standards to ensure, as necessary under the circumstances, record authenticity, integrity, and confidentiality.*
>
> 21 CFR Part 11.30

Well-known hashing algorithms are:

- MD2 and MD5. Both of these algorithms create a 128-bit message digest (RSA Laboratories, RFCs 1319 and 1321, respectively)
- SHA-1 (secure hashing algorithm) is a NIST-sponsored hashing function that has been adopted by the U.S. government as a standard
- Ripe-MD-160 is an algorithm from the European Community. It produces a 160-bit message digest

DATA ENCRYPTION

Encryption refers to the process of scrambling input messages/records, called the *plaintext*, with a user-specified password (password-based encryption algorithm) or key (secret-key algorithm) to generate an encrypted output, called *ciphertext*. It is not possible to recover the original *plaintext* from the *ciphertext* without the user-specified password or key. The algorithms comprising the user-specified password or key, and the *plaintext*, are called ciphers. Encryption is mostly used to protect the privacy of messages or records.

In the 1960s, Horst Feistel designed one of the first modern encryption algorithms at IBM. Until recently all encryption algorithms were based on encrypting and decrypting using the same private key. Only the owner knows the private key and will only share this key with the parties he/she wants to communicate with. The sharing of a secret key introduces an element of risk because compromising the secrecy of the private key may also compromise the integrity of the data.

---

[3] Analogous to a public notary, Digital Notary Services provide trusted date-and-time stamp for a document, so that someone can later prove that the document existed at a point in time. May also verify the signature(s) on a signed document before applying the stamp.

In 1976, Whitfield Diffie developed public-key encryption as an alternative to private-key encryption. Public-key encryption is based on two halves of the same key that are generated with special software at the same time. The key pair are mathematically related so that the private key cannot be determined from the public key. Only one of the two halves of the key pair is required to encrypt a message, with the other half being used for decryption. In public key cryptography, one half of the key pair (the private key) is assigned to an individual, and is closely guarded and securely stored on the user's local disk in either an encrypted format or as part of a token that interfaces with the computer. The other half of the key is published in a public directory where all users can access it and this therefore referred to as the public key.

Public-key cryptography, when properly implemented and used, enables people to communicate in secrecy, and to sign documents, with almost absolute security and without ever having to exchange a private key.

Provided there is a strong linkage between the owner and the owner's public key, the identity of the originator of a message or record can be traced to the owner of the private key. Public-key encryption can play an important role in providing the necessary security services including confidentiality, authentication, integrity of records, and digital signatures.

PKCS is a family of standards for public-key encryption developed by RSA Laboratories.[4] They describe the syntax for several data structures used with public-key cryptography.

One well-known product based on PKCS is the PKI. PKI is a combination of software, encryption technologies, server platforms, workstations, policies, and services used for administering certificates and public-private key pairs.

It enables organizations to protect the security of their communications and business transactions when using networks. PKI is used to secure e-mails, Web browsers, VPNs, and end applications.

In a traditional PKI architecture, a CA is a trusted party that vouches for the authenticity for the entity in question. The CA notarizes public keys by digitally signing the certificates using the CA's private key, which is linked to the entities[5] concerned.

A Certificate Server is a repository for digital certificates. End applications that are PKI-enabled verify the validity and access privileges of a certificate by checking the certificate's profile status, which is protected in the repository.

The Security Server provides services for managing users, digital certificate security policies, and trust relationships in a PKI environment.

PKI architectures can be classified as one of three configurations: a single CA, a hierarchy of CAs, or a mesh of CAs. Each of the configurations is determined by the fundamental attributes of the PKI: the number of CAs in the PKI, where users of the PKI place their trust, and the trust relationships between CAs within a multi-CA PKI.

Digital certificates are digitally signed data structures that contain information such as an entity's name, public key, signature algorithm, and extensions. This information resides in the Active Directory™ located on the Certificate Server. The International Telecommunications Union X.509 standard is the most widely used digital certificate specification. A sample X.509v3 compliant digital certificate[6] data structure can be found in Sidebar 24-1. The

---

[4] RSA Laboratories, in collaboration with Apple, Digital, Lotus, Microsoft, MIT, Northern Telecom, Novell and Sun, developed a family of standards describing data structures used with public key cryptography.

[5] An entity can be a person, server, organization, account, or site.

extensions can be used to tailor digital certificates to meet the needs of the end applications.

End applications either have to be PKI-enabled, PKI-aware out of the box, or they have to be enabled separately. The enabling process may involve using PKI-vendor 'plug-ins' (e.g., Entrust Technologies and Shym Technologies have 'plug-ins' for SAP) that can be added into the end application, or it may involve far more detailed programming.

With respect to component-level PKI interoperability, the contract developer/integrator must understand that enabling an end application to operate with one vendor's PKI products does not necessarily ensure that the end application will also operate with a different vendor's PKI products. If companies do not want to find themselves stranded on their own PKI-

```
Version #
Serial #
Signature Algorithm
Issuer Name
Validity Period
Subject Name
Subject Public Key
Issuer Unique ID
Subject Unique ID
Extensions

Digital Signature
```

Sidebar 24–1. Sample X.509.

island, then they must plan to integrate with other installations. However, if a company makes their installation capable of accepting x.509v3 compliant digital certificates, this does provide interoperability. The end application can accept such certificates from multiple vendors CAs, provided that the certificates honour a consistent certificate profile for their extension fields. A sample application-programming interface to a PKI service can be found at the following address: http://csrc.nist.gov/pki/pkiapi/welcome.htm.

The issue of PKI interoperability becomes complicated when it is compared with inter-domains. Interdomain interoperability involves several technologies and policy-related challenges. Explaining these challenges is, however, beyond the scope of this book.

Some encryption schemes include:

- Rivest, Shamir, and Aldeman (RSA), 1977.
- Diffie-Hellman
- ElGamal Public Key system
- Digital Signature Standard (DSS) that uses the DSA
- RC4 used by Microsoft® Kerberos (128-bit key length)

The FIBS approved algorithm includes:

- DES
- Triple DES, as specified in ANSI X9.52, Triple Data Encryption Algorithm Modes of Operation
- Skipjack

---

[6] X.509 Digital Certificate is the International Telecommunications Union — Telecommunication Standardization Section (ITU-T) recommendation that defines a framework for the provision of authentication services under a central control paradigm represented by a 'Directory.' The recommendation describes two levels: **simple authentication**, using a password as verification of claimed identity, and **strong authentication**, involving credentials formed by using cryptographic techniques, the 'certificate.' The format of the certificate structure is defined along with responsibilities of the CA with regard to establishing and maintaining trust.

Table 24–1. Encryption Strength.

Key Length (bits)	Amount of Time to Break
30	N/A. Can be cracked by 'brute force' guessed using a powerful PC
40	4 hours
56	22 hours (1999 figure)
64	Are probably breakable by powerful computers
80	Less than 7 months. Read the note for 512-bit key length
128	Less than 7 months. Read the note for 512-bit key length
128 AES	149 trillion years. Read the note on the new encryption standard below
512	7 months (1999 figure, refer to the Internet address http://www.cwi.nl/~kik/persb-UK.html)
1024	28,000 billion years (1996 figure. Using search techniques, with a 100 MIPS (Million Instructions Per Second) computer – equivalent to a 200-MHz Pentium)

The key length, in bits, determines the encryption strength of the cryptographic algorithm. Refer to Table 24–1. For example, DES, which was adopted in 1977, uses a 56-bit key length and can be 'cracked' by specialized computers in only a few hours. In addition, the use of DES is not recommended due to the susceptibility to cryptographic exhaustion attack. One replacement for DES, is Triple DES which uses two 56-bit keys.

Another example of the lack of strength of the cryptographic algorithms is related to the plain text. A 40-bit encryption is the default for browsers and this can be easily removed using an old desktop PC in four hours.

The U.S. government will be upgrading its data-encryption standard. The new AES, which was selected after a four-year study, supports key sizes of 128, 192, and 256 bits. According to NIST, if a theoretical machine were built fast enough to crack a standard DES in one second, it would still take the same machine 149 trillion years to break a 128-bit AES key.

Advances in the private sector include the XTR crypto-system. This system makes the current schemes into more efficient and compact implementations.

The US government controls the export of cryptographic implementations. A recent amendment to the Export Administration Regulations (15 CFR Parts 734, 740, 742, 770, and 774) eases the restrictions on encrypted applications. The rules governing export can be quite complex, since they consider multiple factors. In addition, encryption is a rapidly changing field, and the rules may change from time to time. Questions concerning the export of a particular implementation in the USA should be addressed to the appropriate legal counsel.

DIGITAL SIGNATURES

Refer to Chapter 23.

WINDOWS® OS

The Microsoft Windows® 2000 is the first operating system to build PKI (adhering to PKI standards) into its core, and allows the system programmers to establish and maintain cryptographic-based security infrastructure and the foundation for a secure network. Using

Windows® 2000 with other popular Microsoft® applications allows users to gain the following security capabilities:

- secure e-mail using Outlook®
- secure Web access using Internet Explorer, Microsoft's IIS, and Windows® 2000 servers. The RSA algorithm is included as part of the Web browser by Microsoft
- file encryption using Windows® 2000 EFS
- smartcard based single sign-on
- VPNs using the IPSec/VPN capabilities within Windows® 2000

The primary components in Windows® 2000 that support PKI environments are Certificate Services, Active Directory™, standards-based PKI-enabled applications, and Exchange Key Management Services.

Certificate Services is a core operating system service, which allows businesses to act as their own CA. Based on the organization's CA approval instructions, the CA can issue and manage a digital certificate to represent its e-business identities. Windows® 2000 supports multiple levels of a CA hierarchy and cross-certification as well as off-line and on-line CA for maximum flexibility. A CA can issue digital certificates for purposes such as digital signatures, secure e-mail, and the authentication of Web servers using SSL[7] or TLS.

PKCS 1, 7, and 10 are used by the Windows® 2000 Certificate Services. For example, PKCS 10 describes how to construct a certificate request message. After the Windows® 2000 Certificate Services processes the request, the operating system will issue a x.509v3-compliant digital certificate that accepts or rejects the request. PKCS 10 may be used to implement access controls in networked and end applications.

Items such as file permissions, registry settings, password usage, user rights, and other issues associated with Windows® 2000 security have a direct impact on the Certificate Services security.

The Active Directory™ provides information from authoritative sources about people and resources (such as employees, partners, customers, servers, roles, directories, information, digital certificates, and so forth). It serves as the internal and external Certificate Distribution System. Active Directory™ is also the centralized management interface for digital certificate issuance.

Windows® 2000 introduces the concept of the enterprise certificate authority. This feature is integrated with the Active Directory™ and provides other features such as SSL client mapping and smartcard logon.

Some of the standards based PKI-enabled application include: Internet Explorer, Encrypting File System, IPSec, Outlook®, and Outlook Express.

The KMS is the component of Microsoft® Exchange that allows the archiving and retrieval of keys used to encrypt e-mail. Both Exchange Servers 5.5 and 2000 integrate with a Windows® 2000 CA for the issuance of x.509v3-compliant digital certificates.

The National Security Agency (refer to internet address: www.nsa.gov) has developed security configuration guidance for Windows® 2000 in order to provide information on the services that are available in the Microsoft® Windows® 2000 environment and how to integrate these services into their network architecture.

---

[7] Secure Sockets Layer (SSL) is considered to be the industry-standard protocol for secure, Web-based communications. A recent version includes data encryption between the server and the browser and its support client authentication.

Chapter 25

All Together

Following the ISO/IEC International Standard,[1] primary SLC processes can be used to guide the reader on how to put together all the elements discussed in this book. It is not the intention of this chapter to develop a paradigm or model for the regulated industry.

Primary SLC processes depend on who initiates or performs the development, operation, or maintenance of software products. These primary parties are the acquirer, the supplier, the developer, the operator, and the maintainer of software products. The primary SLC processes consist of acquisition process, supply process, development process, operation process, and maintenance process.

The following defines each process and the activities in this book associated with the process.

ACQUISITION PROCESS

> Defines the activities of the acquirer, the organization that acquires a system, software product, or service.

The activities in this book associated with the Acquisition Process include:

Initiation
Chapter 8
•   Acquisition Planning

Request for Proposal
Chapter 7
•   Requirements gathering, including technical constraints
Attachment E
•   Requirements gathering and specification

Contract Preparation Update
This activity is beyond the scope of this book.

Supplier Monitoring
Chapter 9
•   In-process audits
•   Code reviews, inspections, audits

---

[1]   ISO1207:1995, *Information technology — Software life cycle processes.*

Acceptance and Completion
Appendix E
• FAT
• SAT

SUPPLY PROCESS

> Defines the activities of the supplier, the organization that provides the system, software product, or software service to the acquirer.

The activities in this book associated with the Supply Process include:

Initiation; Preparation of Response; Contract
These activities are beyond the scope of this book.

Planning
Chapter 8
• Development Planning

Execution and Control
Develop product based on Development Process
Maintain product based on Maintenance Process

Review and Evaluation
Chapter 9
• In-process audits
• Code reviews, inspections, audits
Appendix E
• FAT
• SAT

Delivery and Completion
Appendix E
• Formal release for use

DEVELOPMENT PROCESS

> Defines the activities of the developer, the organization that defines an develops the software product.

The activities in this book associated with the Development Process include:

Process Implementation
Chapter 5
• Computer Validation Management Cycle

Chapter 6
• Computer Validation Program Organization
Chapter 7
• Computer Systems Validation Process
Chapter 8
• Development Planning
Chapter 11
• SLC Documentation
Appendix E
• Sample Development Activities

Requirements Analysis
Chapter 7
• Conceptualization
• Development
  – Requirements gathering
    ➤ Part 11, Chapter 20
    ➤ Security
    ➤ Operation
    ➤ Safety
  – System Specification
Chapter 22
• Electronic records storage

Design Analysis
Chapter 7
• Development
  – Technical design

Coding and Testing
Chapter 7
• Development
  – Program build
  – Chapter 9

Integration
Chapter 9
• Program built testing
Appendix E
• Conduct Software Module (Unit) Testing and Integration Testing
• Hardware Integration and Program Build

System Testing
Appendix E
• FAT
Chapter 9
• In-process audits
• Code reviews, inspections, audits

Installation
Chapter 10
* Hardware Installation Qualification
* Software Installation Qualification

Acceptance Support
Appendix E
* SAT

OPERATION PROCESS

> Defines the activities of the operator, the organization that provides the service of operating a computer system in its live environment for its users.

The activities in this book associated with the Operation Process include:

Process Implementation and User Support
Chapter 12 and Appendix F
* Procedural controls applicable to operation
Chapter 13
* Change Control
Appendix E
* Computer Systems Incidents

Operational Testing
Appendix E
* Support to OQs and PQs
* Support computer system release activities

System Operation
Chapter 18
* Business Continuity
* Operational Management
* Periodic reviews

MAINTENANCE PROCESS

> Defines the activities of the maintainer, the organization that provides the service of managing modifications to the software product to keep it current and in operational fitness. This process includes the migration and retirement of the software product.

The activities in this book associated with the Maintenance Process include:

Process Implementation
Chapter 8
• Maintenance Project Planning
Chapter 13
• Change Control

Problem and Modification Analysis; Modification Implementation; Maintenance Review/Acceptance
The maintenance activities and work products should be consistent with the Development Process above.

Migration and Software Retirement
Attachment E
• Retirement
Chapter 22
• Records Preservation Strategies

Chapter 26

The Future

A recent paper[1] addressed the future of computer validation. The authors of this paper anticipated how the current industry events and trends would affect computer validation.

Technology is one of many trends constantly affecting computer validation. One area requiring special attention is the introduction of expert systems into the FDA-regulated environment. The simplest form of artificial intelligence generally used in applications (such as mortgages, credit card authorization, fraud detection, e-commerce, personalization) is the rule-based system, also known as the expert system (ES).

An expert system is a way of encoding a human expert's knowledge of a fairly narrow area into a computer system. There are a couple of advantages to doing so. One is that the human expert's knowledge becomes available to a large range of people. Another advantage is that if you can capture the expertise of an expert in a field, then that knowledge is not lost when he/she retires or leaves the firm.

Expert systems differ from standard procedural or object-oriented programs in that there is no clear order in which code executes. Instead, the knowledge of the expert is captured in a set of rules, each of which encodes a small piece of the expert's knowledge. Each rule has a left-hand side and a right-hand side.

The left-hand side contains information about certain facts and objects that must be true in order for the rule to potentially be executed. Any rules whose left-hand sides match in this manner at a given time are placed on an agenda.

One of the rules on the agenda is picked (there is no way to predict which one), and its right-hand side is executed, and it is removed from the agenda. The agenda is then updated (generally using a special algorithm called the Rete algorithm), and a new rule is picked to execute. This continues until there are no more rules on the agenda.

Because the actions of the rule-based system tend to be hidden from view, people tend not to realize just how extensively they are used.

A typical rule for a mortgage application might look something like this:

IF	(number-of-30-day-delinquencies > 4)
AND	(number-of-30-day-delinquencies < 8)
THEN	increase mortgage rate by 1%

As the reader can see, a rule bears a close resemblance to an if-then-else statement, but unlike an if-then-else statement, it stands alone and does not fire in any predetermined order relative to other if-then-else statements.

In a way, a rule-based system might almost be thought of as being similar to a multi-threaded system in that just as one doesn't know which thread will execute next, one doesn't know which rule will execute next. However, rule-based systems are generally implemented as single-thread programs.

---

[1] R. Buihandojo, D. J. Bergeson, B. Bradley, P. D'Eramo, L. Huber, O. López et al. The Future State of Computer Validation, *Pharmaceutical Technology*, July/September 2001.

The advantage to this type of approach, as opposed to a procedural approach, is that if the system is well designed, then the expert's knowledge can be maintained fairly easily, just by altering whichever rules need to be altered. Definitely, many rule-based systems come along with a rules editor that allows for rules to be easily maintained by nontechnical people.

Rules are generally implemented in something called a rules engine, which provides a basic framework for writing rules and then for running them in the manner described above.

In a standard procedural or object-oriented environment, sequencing is a key element in the validation of a computer system. Sequencing encompasses instruction to operators, operation/process sequencing, and Part 11-related operational checks. So, how are these systems to be tested? Who will make sure that the calling sequence is correct?

Another area with immediate impact to computer validation is the current federal legislation, Electronic Signatures in Global and National Commerce Act, on definitions of electronic records. This definition is not consistent with Part 11. The current federal legislation defines electronic records in more general terms than Part 11 does. The primary element into consideration in the current federal legislation is whether the data is accessible once put in storage, not considering the technology used. This definition will consider data in transient memory as electronic records. One key element impacting validation is the regulatory requirements to data stored in transient memory, including audit trails.

Particularly for global companies, harmonization of SLC and development methodology practices is another area that should be evaluated. To achieve this, a common international standard may be used as a framework. The two (2) key international standards are ISO9000–3[2] and ISO12207[3]. Implementation of these standards can be found in local standard bodies. The evolution of GAMP[4] incorporates these two standards. Refer to Figure 26–1. The framework of 12207 includes five 'primary processes': acquired, supplying, developing, operating, and maintaining software. It divides the five processes into 'activities,' and the activities into 'tasks,' while placing requirements upon their execution. It also specifies eight 'supporting processes': documentation, configuration management, quality assurance, verification, validation, joint review, audit, and problem resolution — as well as four 'organizational processes' — management, infrastructure, improvement, and training. Now that ISO9001:1994, ISO9002:1994 and ISO9003:1994 have been integrated into the new ISO9001:2000, ISO9000–3 may be reviewed for consistency with the ISO standards.

Specifically for medical devices, the Medical Device Software Committee of the Association for the Advancement of Medical Instrumentation (AAMI) has taken on the task of developing a standard[4] for use in the medical device business sector based upon the framework established in ISO12207.

In the future, computer systems validation needs to be adaptable to new technologies and regulation changes. In the future, the industry might be more experienced managing computer technology suppliers and contract developers. The cost of implementing and maintenance of computer systems can be streamlined by effectively leveraging the work of computer technology suppliers and contract developers. The utilization of contract developers verifications and testing to demonstrate that an automated solution meets the defined specification is one of many activities to streamline the project costs.

---

[2]   ISO9000–3:1997, '*Quality management and quality assurance standards Part 3, Guidelines for the application of ISO9001: 1994 to the development, supply, installation and maintenance of computer software.*' Author's note: At the time of this book publishing, this guideline is in the process of being updated to align with ISO9001:2000.

[3]   ISO12207:1995, 'Information technology — Software life cycle processes.'

[4]   ANSI/AAMI SW68:2001, '*Medical device software — Software life cycle processes.*'

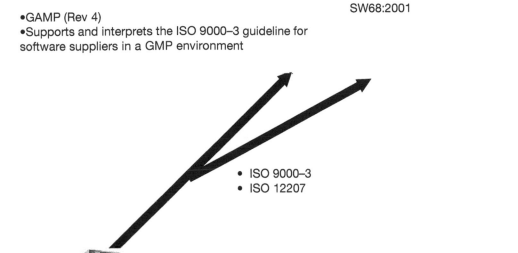

•ANSI/AAMI
SW68:2001

•GAMP (Rev 4)
•Supports and interprets the ISO 9000–3 guideline for
software suppliers in a GMP environment

• ISO 9000–3
• ISO 12207

ISO 9001:1994 – Quality Management and Quality Assurance
Standards

Figure 26–1. ISO 9001 Evolution.

Glossary of Terms

For additional terms, refer to the *Glossary of Computerized System and Software Development Terminology*[1] and *A Globally Harmonized Glossary of Terms for Communicating Computer Validation Key Practices*.[2]

Abstraction This is a basic principle of software engineering, and enables an understanding of an application and its design, and the management of complexity.

Acceptance Criteria The criteria that a system or component must satisfy to be accepted by a user, customer, or other authorized entity (IEEE).

Acceptance Test Testing conducted to determine whether a system satisfies its acceptance criteria and to enable the customer to determine whether to accept the system (IEEE).

Access The ability or opportunity to gain knowledge of stored information (DOD 5015.2–STD).

Application A group of computer instructions that is used to accomplish a business function, control a process, or facilitate decision-making. Refer to **Application Software** in *Glossary of Computerized System and Software Development Terminology*, August 1995.

Approver(s) In the context of configuration management, the approver is the person(s) responsible for evaluating the recommendations of the reviewers of deliverable documentation, and for making a decision on whether to proceed with a proposed change, and for initiating the change request implementation.

Auditor In the context of configuration management, the auditor is the person responsible for reviewing the steps taken during a development or change management process, to ensure that the appropriate procedures have been followed.

Audit Trail An electronic means of auditing the interactions between records within an electronic system, so that any access to the system can be documented as it occurs, to identify unauthorized actions in relation to the records, e.g., modification, deletion, or addition (DOD 5015.2–STD).

Authentication The process used to confirm the identity of a person, or to prove the integrity of specific information. In the case of a message, authentication involves determining the message source, and providing assurance that the message has not been modified or replaced in transit (ABA).

Authenticity A condition that proves a record is authentic and/or genuine, based on its mode (the method by which a record is communicated over space or time); on its form (format and/or media that a record takes when received); state of transmission (the primitiveness,

---

[1] FDA, *Glossary of Computerized System and Software Development Terminology*, Division of Field Investigations, Office of Regional Operations, Office of Regulatory Affairs, Food and Drug Administration, August 1995.

[2] Herr, Robert R. and Wyrick, Michael L., A Globally Harmonized Glossary of Terms for Communicating Computer Validation Key Practices, *PDA Journal of Pharmaceutical Science and Technology*, March/April 1999.

completeness, and effectiveness of a record when initially set aside after being made or received); and manner of preservation and custody (DOD 5015.2–STD).

Automated systems Include a broad range of systems including, but not limited to, automated manufacturing equipment, automated laboratory equipment, process control, manufacturing execution, clinical trials data management, and document management systems. The automated system consists of the hardware, software, and network components, together with the controlled functions and associated documentation (GAMP).

Certification authority The authority (part of the public-key infrastructure) in a network that issues and manages security credentials and public key for message encryption and decryption from a certificate server (NARA).

Change Any variation or alteration in form, state, or quality. It includes additions, deletions, or modifications impacting the hardware or software components used affecting operational integrity, service level agreements, or the validated status of applications on the system.

Cipher Series of transformations that converts plaintext to ciphertext using the Cipher key.

Cipher key Secret cryptography key that is used by the Key Expansion routine to generate a set of round keys.

Cipher text Data output from the Cipher or input to the Inverse Cipher.

Code audit An independent review of source code by a person, team, or tool to verify compliance with software design documentation and programming standards. Correctness and efficiency may also be evaluated (IEEE).

Code inspection A manual (formal) testing (error detection) technique where the programmer reads source code, statement by statement, to a body that can ask questions; for example, when analyzing the program logic, analyzing the code with respect to a checklist of historically common programming errors, and analyzing its compliance with coding standards. This technique can also be applied to other software and configuration items (Myers/NBS).

Code review A meeting at which software code is presented to project personnel, managers, users, customers, or other interested parties for comment or approval (IEEE).

Code walkthrough A manual testing (error detection) technique where program (source code) logic (structure) is traced manually (mentally) by a group or body with a small set of test cases; the state of program variables is simultaneously manually monitored, to analyze the programmer's logic and assumptions (Myers/NBS).

Complexity In the context of this book, complexity means the degree to which a system or component has a design or implementation that is difficult to understand and verify. Sample factors for establishing the complexity of a system or a component can be reviewed in Appendix D.

Compliance Compliance covers the adherence to application-related standards or conventions or regulations in laws and similar prescriptions.

Configurable software Application software, sometimes general purpose, written for a variety of industries or users in a manner that permits users to modify the program to meet their individual needs (FDA).

Computer A functional unit that can perform substantial computations, including numerous arithmetic operations and logical operations without human intervention.

Computer System (1) A system including the input of data, electronic processing, and the output of information to be used either for reporting or automatic control (EU PIC/S). (2) A functional unit, consisting of one or more computers and associated peripheral input and output devices, and associated software, that uses common storage for all or part of a program and also for all or part of the data necessary for the execution of the program; executes user-written or user-designated programs; performs user-designated data manipulation, including arithmetic operations and logic operations; and execute programs that modify themselves during their execution. A computer system may be a standalone unit or may consist of several interconnected units (ANSI).

Computer Systems Validation (CSV) The formal assessment and reporting of quality and performance measures for all the life-cycle stages of software and system development, its implementation, qualification and acceptance, operation, modification, requalification, maintenance, and retirement, such that the user has a high level of confidence in the integrity of both the processes executed within the controlling computer system(s), and in those processes controlled by and/or linked to the computer system(s), within the prescribed operating environment(s) (MCA).

Concurrent Validation In some cases, a drug product or medical device may be manufactured individually or on a one-time basis. The concept of prospective or retrospective validation as it relates to those situations may have limited applicability. The data obtained during the manufacturing and assembly process may be used in conjunction with product testing, to demonstrate that the instant run yielded a finished product meeting all of its specifications and quality characteristics (FDA).

Control System Included in this classification are Supervisory Control and Data Acquisition Systems (SCADA), Distributed Control Systems (DCS), Statistical Process Control systems (SPC), Programmable Logic Controllers (PLCs), intelligent electronic devices, and computer systems that control manufacturing equipment or receive data directly from manufacturing equipment PLCs.

Criticality In the context of this book, criticality means the regulatory impact to a system or component. Sample factors that establish the criticality of a system or a component can be reviewed in Appendix D.

Custom-built software Also known as a Bespoke System, Custom-Built Software is software produced for a customer, specifically to order, to meet a defined set of user requirements (GAMP).

Digital certificate A credential issued by a trusted authority. An entity can present a digital certificate to prove its identity or its right to access information. It links a public-key value to information that identifies the entity, associated with the use of the corresponding private key. Certificates are authenticated, issued, and managed by a trusted third party called a CA.

Documentation Manuals, written procedures or policies, records, or reports that provide information concerning the use, maintenance, or validation of a process or system involving either hardware or software. This material may be presented from electronic media. Documents include, but are not limited to, Standard Operating Procedures (SOPs), Technical Operating Procedures (TOPs), manuals, logs, system development documents, test plans, scripts and results, plans, protocols, and reports. Refer to **Documentation** and **Documentation, level of** in the *Glossary of Computerized System and Software Development Terminology*, August 1995.

Emergency change A change to a validated system determined necessary to eliminate an error condition that prevents the use of the system and interrupts business function.

Emulation The process of mimicking, in software, a piece of hardware or software so that other processes 'think' that the original equipment/ function is still available in its original form. Emulation is essentially a way of preserving the functionality of, and access to, digital information that might otherwise be lost due to technological obsolescence.

Encryption The process of converting information into a code or cipher. A secret key, or password, is required to decrypt (decode) the information, which would otherwise be unreadable.

Entity A software or hardware product that can be individually qualified or validated.

Establish Establish is defined in this book as meaning to define, document, and implement.

Factory Acceptance Test (FAT) An acceptance test in the supplier's factory, usually involving the customer (IEEE).

Field Devices Hardware devices typically located in the field at or near the process, necessary for bringing information to the computer or implementing a computer-driven control action. Devices include sensors, analytical instruments, transducers, and valves.

GxP A global abbreviation intended to cover GMP, GCP, GLP, and other regulated applications in context.

GxP Computerized Systems A computerized system that performs regulated operations that are required to be formally controlled under the Federal Food, Drug, and Cosmetic Act, the Public Health Service, and/or an applicable regulation.

Hybrid systems In the context of Part 11, hybrid computer systems save required data from a regulated operation to magnetic media. However, the electronic records are not electronically signed. In summary, in hybrid systems some portions of a record are paper and some are electronic.

Impact of change The impact of change is the effect of change on the GxP computerized system. The components by which the impact of change is evaluated may include, but not be limited to, business considerations, resource requirements and availability, application of appropriate regulatory agency requirements, and criticality of the system.

Inspection A manual testing technique in which program documents [specifications (requirements, design), source code or user's manuals] are examined in a very formal and disciplined manner to discover any errors, violations of standards or other problems. Checklists are typical vehicles used in accomplishing this process. Refer to Chapter 9.

Installation qualification Establishing confidence that process equipment and ancillary systems are capable of consistently operating within established limits and tolerances (FDA).

Key practices Processes essential for computer validation that consists of tools, workflow, and people (PDA).

Legacy systems Production computer systems that are operating on older computer hardware or are based on older software applications. In some cases, the vendor may no longer support the hardware or software.

Logically secure and controlled environment A computing environment, controlled by policies, procedures, and technology, which deters direct or remote unauthorized access that could damage computer components, production applications, and/or data.

Major change A change to a validated system that is determined by reviewers to require the execution of extensive validation activities.

Metadata Data describing stored data; that is, data describing the structure, data elements, interrelationships, and other characteristics of electronic records (DOD 5015.2–STD).

Migration Periodic transfer of digital materials from one hardware/software configuration to another, or from one generation of computer technology to a subsequent generation.

Minor change A change to a validated system that is determined by reviewers to require the execution of only targeted qualification and validation activities.

Model A model is an abstract representation of a given object.

Module testing Refer to **Testing, Unit** in the *Glossary of Computerized System and Software Development Terminology*, August 1995.

NEMA enclosure Hardware enclosure (usually a cabinet) that provides different levels of mechanical and environmental protection to the devices installed within it.

Noncustom Purchased Software Package A generally available, marketed software product that performs specific data collection, manipulation, output, or archiving functions. Refer to **Configurable, off-the-shelf software** in the *Glossary of Computerized System and Software Development Terminology*, August 1995.

Operating environment All outside influences that interface with the computer system (GAMP).

Ongoing evaluation A term used to describe the dynamic process employed after a system's initial validation that can assist in maintaining a computer system in a validated state.

Operational testing Refer to **Operational Qualification** in the *Glossary of Computerized System and Software Development Terminology*, August 1995.

Operating system Software controlling the execution of programs and providing services such as resource allocation, scheduling, input/output control, and data management. Usually, operating systems are predominantly software, but partial or complete hardware implementations are possible (ISO).

Password A character string used to authenticate an identity. Knowledge of the password that is associated with a userID is considered proof of authorization to use the capabilities associated with that userID (CSC-STD-002–85).

Packaged software Software provided and maintained by a vendor/supplier that can provide general business functionality or system services. Refer to **Configurable, off-the-shelf software** in the *Glossary of Computerized System and Software Development Terminology*, August 1995.

Periodic review A documented assessment of the documentation, procedures, records, and performance of a computer system to determine whether it is still in a validated state and what actions, if any, are necessary to restore its validated state (PDA).

Personal Identification Number A PIN is an alphanumeric code or password used to authenticate the identity of an individual.

Physical environment The physical environment of a computer system comprising the physical location and the environmental parameters in which the system physically functions.

Planned change An intentional change to a validated system for which an implementation and evaluation program is predetermined.

Policy A directive that usually specifies what is to be accomplished.

Predicate regulations Federal Food, Drug, and Cosmetic Act, the Public Health Service Act, or any FDA Regulation, with the exception of 21 CFR Part 11. Predicate regulations address the research, production, and control of FDA regulated articles.

Procedural controls (1) Measures taken to ensure the trustworthiness of records and signatures established through the implementation of standard operating procedures (SOPs). (2) Written and approved procedures providing appropriate instructions for each aspect of the development, operations, maintenance, and security applicable to computer technologies. In the context of regulated operations, procedural controls should have QA/QC controls equivalent to the applicable predicate regulations.

Process system The combination of the process equipment, support systems (such as utilities), and procedures used to execute a process.

Production environment The operational environment in which the system is being used for its intended purpose, i.e., not in a test or development environment.

Production verification (PV) Documented verification that the integrated system performs as intended in its production environment. PV is the execution of selected Performance Qualification (PQ) tests in the production environment using production data.

Prospective validation Validation conducted prior to the distribution of either a new product, or product made under a revised manufacturing process, where the revisions may affect the product's characteristics. (FDA)

Qualification (1) Action of proving that any equipment works correctly and actually leads to the expected results. The word *validation* is sometimes widened to incorporate the concept of qualification (EU PIC/S). (2) Qualification is the process of demonstrating whether a computer system and associated controlled process/operation, procedural controls, and documentation are capable of fulfilling specified requirements.

Qualification protocol A prospective experimental plan stating how qualification will be conducted, (including test parameters, product characteristics, production equipment, etc.) and decision points on what constitutes an acceptable test. When executed, a protocol is expected to produce documented evidence that a system or subsystem performs as required.

Qualification reports These are test reports that evaluate the conduct and results of the qualification carried out on a computer system.

Raw data The original data that has not been manipulated or data that cannot be easily derived or recalculated from other information.

Record A record consists of information, regardless of medium, detailing the transaction of business. Records include all books, papers, maps, photographs, machine-readable materials, and other documentary materials, regardless of physical form or characteristics, made or received by an Agency of the United States Government under Federal law; or in connection with the transaction of public business and preserved or appropriate for preservation by that Agency or its legitimate successor as evidence of the organization, functions, policies, decisions, procedures, operations, or other activities of the Government or because of the value of data in the record (44 U.S.C. 3301, reference (bb)).

Record owner 'Record owner' means a person or organization who can determine the contents and use of the data collected, stored, processed or disseminated by that party regardless of whether or not the data was acquired from another owner or collected directly from the provider.

Record reliability A reliable record provides contents that can be trusted as a full and accurate representation of the transactions, activities, or facts to which they attest, and can be depended upon in the course of subsequent transactions or activities (NARA).

Regulated operations Process/business operations carried out on an FDA-regulated product covered in a predicated rule.

Replacement The implementation of a new Part 11-compliant system after the retirement of an existing system.

Requalification Repetition of the qualification process or a specific portion thereof.

Remediate software hardware and/or procedural changes employed to bring a system into compliance with 21 CFR Part 11.

Retirement phase The period in the SLC in which plans are made and executed to decommission or remove a computer technology from operational use.

Retrospective evaluation Establishing documented evidence that a system does what it purports to do, based on an analysis of historical information. The process of evaluating a computer system currently in operation against standard validation practices and procedures. The evaluation determines the reliability, accuracy, and completeness of a system.

Retrospective validation See Retrospective evaluation.

Site Acceptance Test (SAT) An acceptance test at the customer's site, usually involving the customer (IEEE).

Software development standards Written policies or procedures that describe practices a programmer or software developer should follow in creating, debugging, and verifying software.

Source Code The human readable version of the list of instructions (programs) that enable a computer to perform a task.

Specification A document that specifies, in a complete, precise, and verifiable manner, the requirements, design, behaviour, or other characteristics of a system or component; and often the procedures for determining whether such provisions have been satisfied (IEEE).

Static analysis (1) Analysis of a program performed without executing the program (NBS). (2) The process of evaluating a system or component based on its form, structure, content, documentation (IEEE).

Standard instrument software Software driven by non-user-programmable firmware, which is configurable (GAMP).

Standard software packages A complete and documented set of programs supplied to several users for a generic application or function (ISO/IEC 2382–20:1990).

System (1) People, machines, and methods organized to accomplish a set of specific functions (ANSI). (2) A composite entity, at any level of complexity, of personnel, procedures, materials, tools, equipment, facilities, and software. The elements of this composite entity are used together within the intended operational or support environment, to perform a given task or achieve a specific purpose, support, or mission requirement (DOD).

System backup The storage of data and programs on a separate media, and stored separately from the originating system.

System Life Cycle (SLC) The period of time commencing from when the system product is recommended, until the system is no longer available for use or is retired.

System owner The person(s) who have responsibility for the operational system, and bear the ultimate responsibility for ensuring a positive outcome of any regulatory inspection or quality audit of the system.

System retirement The removal of a system from operational usage. The system may be replaced by another system or may be removed without being replaced.

System software See Operating System.

Technical controls Measures taken to ensure the trustworthiness of records and signatures established through the application of computer technologies.

Test report Document presenting test results and other information relevant to a test (ISO/IEC Guide 2:1991).

Test script A detailed set of instructions for execution of the test. This typically includes the following:

- Specific identification of the test
- Prerequisites or dependencies
- Test objective
- Test steps or actions
- Requirements or instructions for capturing data (e.g., screen prints, report printing)
- Pass/fail criteria for the entire script
- Instructions to follow in the event that a nonconformance is encountered
- Test execution date
- Person(s) executing the test
- Review date
- Person reviewing the test results

For each step of the test script, the item tested, the input to that step, and the expected result, are indicated prior to execution of the test. The actual results obtained during these steps of the test are recorded on (or attached to) the test script. Test scripts and results may be managed through computer-based electronic tools. Refer to **Test case** in the *Glossary of Computerized System and Software Development Terminology*, August 1995.

Test Nonconformance A test nonconformance occurs when the actual test result does not equal the expected result or an unexpected event (such as a loss of power) is encountered.

Training plan Documentation describing the training required for an individual based on his or her job title or description.

Training record Documentation (electronic or paper) of the training received by an individual that includes, but is not limited to, the individual's name or identifier; the type of training received; the date the training occurred; the trainer's name or identifier; and an indication of the effectiveness of the training (if applicable).

Trustworthy Reliability, authenticity, integrity, and usability are the characteristics used to describe trustworthy records from a record management perspective (NARA).

Transient memory Memory that must have a constant supply of power or the stored data will be lost.

Unplanned (emergency) change An unanticipated necessary change to a validated system requiring rapid implementation.

User back-up/alternative procedures Procedure describing steps to be taken for the continued recording and control of the raw data in the event of a computer system interruption or failure.

Unit A separately testable element specified in the design of a computer software element. Synonymous to component, module (IEEE).

UserID A sequence of characters that is recognized by the computer and that uniquely identifies one person. The UserID is the first form of identification. UserID is also known as a PIN or identification code.

Validated The term 'validated' is used to indicate a status that designates that a system and/or software is compliant with all regulatory requirements.

Validation Action of proving, in accordance with the principles of Good Manufacturing Practice, that any procedure, process, equipment, material, activity, or system actually leads to the expected results (see also qualification) (EU PIC/S).

Validation coordinator A person or designee responsible for coordinating the validation activities for a specific project or task.

Validation protocol A written plan stating how validation will be conducted, including test parameters, product characteristics, production equipment, and decision points on what constitutes acceptable test results (FDA).

Validation plan A multidisciplinary strategy from which each phase of a validation process is planned, implemented, and documented to ensure that a facility, process, equipment, or system does what it is designed to do. May also be known as a system or software Quality Plan.

Validation Summary Report (VSR) Documents confirming that the entire project planned activities have been completed. On acceptance of the Validation Summary Report, the user releases the system for use, possibly with a requirement that continuing monitoring should take place for a certain time (GAMP).

Verification The process of determining whether or not the products of a given phase of the SLC fulfill the requirements established during the previous phase.

Work products The intended result of activities or processes (PDA).

Worst case A set of conditions encompassing upper and lower processing limits and circumstances, including those within standard operating procedures, which pose the greatest chance of process or product failure when compared to ideal conditions. Such conditions do not necessarily induce product or process failure (FDA).

Appendix B

Abbreviations and Acronyms

ABA	American Bar Association
ANDAs	Abbreviated New Drug Applications
ANSI	American National Standard Institute
CA	Certification Authority
CAPA	Corrective and Preventive Actions
CFR	Code of Federal Regulation
cGMP	current Good Manufacturing Practices
CPG	FDA Compliance Policy Guide
DES	Data Encryption Standard
DRM	Device Master Record
DSA	Digital Signature Algorithm
EFS	Encrypting File System
EU	European Union
EU PIC/S	European Union Pharmaceutical Inspection Cooperation Scheme
FAT	Factory Acceptance Test
FD&C Act	US Food Drug and Cosmetic Act
FDA	US Food and Drug Administration
FR	US Federal Register
GAMP	Good Automated Manufacturing Practices
GCP	Good Clinical Practices
GLP	Good Laboratory Practices
GMPs	United States Good Manufacturing Practices
GxP	A global abbreviation intended to cover GMP, GCP, GLP, and other regulated applications. In context GxP can refer to one specific set of practices or to any combination of the three.
I/Os	Inputs and outputs
IEEE	Institute of Electrical & Electronic Engineers
IIS	Internet Information Services
ISO	International Organization for Standardization
IT	Information Technology
KMS	Key Management Service
LAN	Local Area Network
MCA	UK Medicine Control Agency
MES	Manufacturing Execution System
NARA	National Archives and Records Administration
NBS	National Bureau of Standards
NDAs	New Drug Applications
NEMA	National Electrical Manufacturers Association
NIST	National Institutes of Standards and Technology
OECD	Organization for Economic Cooperation and Development
OSHA	US Occupational Safety & Health Administration

PDA	Parenteral Drug Association
P&ID	Process and Instrumentation Drawings
PIC/S	EU Pharmaceutical Inspection Cooperation Scheme
PIN	Personal Identification Number
PKCS	Public Key Cryptography Standards
PKI	Public-Key Infrastructure
PLC	Programmable Logic Controller
R&D	Research and Development
RFP	Request for Proposal
SAP	Systems, Application and Products
SAS	The Statistical Analysis System licensed by the SAS Institute, Inc.
SAT	Site Acceptance Test
SCADA	Supervisory Control and Data Acquisition
SLC	System Life Cycle
SOPs	Standard Operating Procedures
SQA	Software Quality Assurance
SQE	Software Quality Engineering
SSA	US Social Security Administration
SSL	Secure Sockets Layer
TLS	Transport Layer Security
UK	United Kingdom
UPS	Uninterruptable Power Supply
US	United States
VPN	Virtual Privates Network

Appendix C

Applicability of a Computer Validation Model

This appendix explains how to utilize the Part 11 computer validation model (Figure 4–1) and the procedures necessary to determine the model parameters and to apply the results.

The elements that need to be taken into account during the implementation of a computer system are: (i) the type of system (open or closed), (ii) security functions, audit trails, (iii) the operation controlled by the computer system, and (iv) the technology necessary to support the operation and Part 11. These last two elements are not identical for all systems.

An element beyond the scope of the model is the retention of electronic records, but the model can be used to verify and validate the implementation of the system(s) that will hold these records. A subset of the above requirements are applicable to hybrid computer systems,[1] and the implementation of these requirements have been discussed in other articles.[2,3] The fundamental requirements that Part 11 establishes are as follows:

1. **Open/Closed Systems**
2. **Security**
   a. System security
   b. Electronic signature security
   c. UserID and password
      maintenance
   UserID and password security

   Password assignment

   d. Document controls
   e. Authority, operational, and
      location checks
   f. Records protection
3. **Operational Checks**

4. **Audit Trails**
   a. Audit mechanism
   b. Metadata
   c. Display and reporting
5. **Electronic signatures**
   a. Electronic signature without
      biometric/behavioral
      identification
   b. Electronic signature with
      biometric/behavioral
      identification
   c. Signature manifestation

   d. Signature purpose
   e. Signature binding
6. **Certification to FDA**

When determining which requirements need to be implemented, verified, and tested, the following points need consideration: whether the records are held in transient memory; whether the system is a hybrid system; and whether the system manages electronic signatures. The following sections describe the implementation of the model.

---

[1] In hybrid systems, some portions of a record are paper and some electronic.
[2] López, O, Automated Process Control Systems Verification and Validation, *Pharmaceutical Technology*, September 1997.
[3] López, O, Implementing Software Applications Compliant with 21 CFR Part 11, *Pharmaceutical Technology*, March 2000.

SAMPLE SITUATIONS

The objective of the following sample situations is to demonstrate the applicability of the new computer systems validation model.

First Case: Records Held in Transient Memory

Two examples of this case are (i) the temperature profile from an autoclave, and (ii) standalone PLC-based systems. All I/Os are saved in the controller's transient memory. Neither the inputs (e.g., the operator keyboard) nor the outputs (e.g., the operator display) are retrieved from, or saved to, durable media. If the record has not been saved yet to durable media, it will be vulnerable to accidental or intentional alteration.

Because no records are yet saved to a durable media, Part 11 is not applicable to these computer systems, or to other computer systems with similar configuration. However, Part 11 may be used as the model to validate the software, hardware, and any associated interfaces. In this case, the following elements in Part 11 are applicable:

1. Operational checks
2. System security
3. UserID and password security
4. UserID and password maintenance
5. Password assignment
6. Location checks
7. Authority checks
8. Document control
9. Open/Closed systems
10. Record retention/protection

Operational checks are application dependent. Refer to Chapter 20. The reader may realize that the application-dependent requirements become part of the model through the operational verifications process.

In this first case, system security is associated with preventing the accidental or intentional alteration and corruption of the data to be displayed on the screen, or be used to make a decision to control the operation. To avoid accidental or intentional loss of data, the data collected must be defined, along with the procedures used to collect it, and the means to verify its integrity, accuracy, reliability, and consistency. A failure modes-and-effects analysis (FMEA) is one of many methods used to uncover and solve these factors. For example, to avoid data corruption, an ongoing verification program (Chapter 18) should be implemented.

Logical access controls for computer systems, including *authority checks*, form another security aspect. Authority checks are based on the various roles and responsibilities assigned to individuals known to the system. Computer systems must be designed to make distinctions between the controlled access (a) to the system, (b) to functions in the system, and (c) to input and output devices used by the system.

Location checks, when applicable, enable the application software to determine whether the input being generated by a particular device is appropriate.

Passwords are one of many methods used to authenticate authorized users. As security of computer systems performing regulated operations is so very critical, expectations for activities (such as assignment, security, and maintenance of passwords) are clearly established by the FDA. Before Part 11, the FDA expected[4] that passwords would be:

---

[4] FDA, draft of *Guideline for the Validation of Blood Establishment Computer Systems*, September 28, 1993.

- Periodically changed, not be reassigned and not reused
- Unrecognizable as reflections of the user's personal life
- *Not* shared by other users

The FDA expects written procedures or company policies to be in place covering the use and security of passwords, supported by training on the subject. To ensure that computer systems are secure, password files may be encrypted or otherwise secured so that the password cannot be read by ordinary means. Password file encryption is commonly used in modern operating systems.

In general, the *assignment of passwords and their maintenance* is a combination of written procedures and the technologies used to support them. Sample procedures include password aging, minimum password length, password uniqueness, default password management, and account lockout after a reasonable number of unsuccessful log-in attempts.

The focus of the validation program is usually the quality attributes of the system implementation, change control, and the verification and testing of modifications made to the baseline system. The configuration management of the system, including its *documentation,* is a key area of concern. Specifically, documentation management is extremely critical for the information contained in both the master production records, and in the application.

System documentation comprises records relating to system operation and maintenance, from the high-level design documents to the end-user manuals. The control of system documentation includes an assessment of the documentation that may be affected when a modification to the computer system is suggested. The high-level design documentation provides the input for maintenance; user manuals provide the instructions on how to operate the system.

According to Part 11, a computer system is considered closed if the person responsible for the content of electronic records only allows their access by authorized parties. In addition, electronic records must be monitored through the computer system's software with its required logon, security procedures, and audit trails. For PLC-based systems, it is very common to communicate with remote systems via the telephone system, accessed by modem. A connection such as this can and may be used by the system developer/integrator to perform diagnostics: *this practice must be disallowed* due to its inherent security and configuration management weaknesses. If this type of access is allowed, the modem must have strict access controls and restrictions, including protective software, and access restrictions on such systems be designed to prevent unauthorized modifications.

In this example, the user would record on paper each significant step in the operation. The *retention and protection* of the paper-based record(s) is based on the retention requirements that can be found in the applicable predicate rule.

Second Case: Hybrid Computer Systems

Hybrid computer systems save data required by the existing regulations onto durable media. However, these electronic records are not electronically signed. An example of this configuration is provided by computer systems generating paper-based reports, which are printed and manually signed.

For hybrid systems, the electronic record requirements in Part 11 (Sub-Part B) are applicable. The electronic records are maintained and retained in electronic form for the period established by the predicate regulation that require these records. The elements of Part 11 applicable to hybrid systems are:

1. Audit trails and metadata
2. System security
3. UserID and password security
4. UserID and password maintenance
5. Password assignment
6. Operational checks
7. Authority checks
8. Location checks
9. Document control
10. Open/Closed systems
11. Record retention/protection
12. Signature/record linkage

A key element of such a system is the completeness of the recorded information. As part of the *validation process*, the completeness of computer-generated electronic records is verified.

An *audit trail* is a journal recording any modifications carried out by the users, or by the processes operating on behalf of the user, to the electronic records. This mechanism provides the capability to reconstruct modified data and must not obscure the previously entered data. The tracking mechanism includes adding computer-generated date and time stamps to the record, to indicate when the record was modified, the types of modifications performed, and the identity of the person performing the modification to the record.

The implicit requirements in Part 11 include the availability of the tools necessary to display and/or print audit trail records, and associated metadata, in human readable form. Another important use of audit trails is the protection from subsequent unauthorized alteration and destruction, and to permit the detection of security violations and after-the-fact investigations into their cause.

The point at which data becomes an electronic record should reflect its origin and intended purpose. For manufacturing systems, this may be the initial point of data acquisition and this point would also form the basis for the recording of audit trails for changes to a record.

Third parties can provide custom code, which can be integrated within most common database systems in order to implement audit trails. This approach provides a technological foundation for the proper administration of electronic records, without replacing the application that manages them.

For example, audit trails may be implemented by using database triggers. A trigger is a feature of the Oracle™ database, which allows code written by the user to be executed based on an event. The concept implies that the edition of a record will produce an event. The event may be defined as an insert, update, or deletion of data from a database table.

An alternative method is to replace the complete application with an 'out-of-the-box' application, which includes an audit-trail feature.

Another alternative is to prevent modifications to the records at the data collection and acquisition domains. The permanent repository system may be located at the business logistic systems level.[5] At this level, the electronic records are either maintained while in active use, or are maintained off-line when their use is less frequent. This setting must be evaluated and, preferably, documented as part of a company wide strategy. The selected settings require procedural and technological controls.

Data is most often transferred using available drivers. The transfer/transcription process, from the secondary repository to the permanent repository, shall be completely qualified. The qualification must include all of the appropriate documentation and supporting data. The objective of the qualification must be to ensure the accurate transfer of electronic records and their associated content, context, and structure.

---

[5] Enterprise/Control Integration standard, ISA SP95.01. The plant management level is the plan repository area (e.g., database server(s)) that unifies production, quality control, inventory, and warehouse data.

An important item to be addressed during the implementation of every computer system is the long-term archiving of data and any associated metadata. The hardware and software supporting the repository must be robust, but also flexible enough to withstand the regulatory electronic records retention requirements.

The use of audit trails is not restricted to the modifications of the records. Audit trails may also be used to record operator entries and actions during the operation of the computer system, which can be recorded without the application of electronic signatures.

In addition to security elements referenced in the previous case, additional *security* functions are needed to cover electronic records. External safeguards must be in place to ensure that access to the computer system and associated data is restricted to authorized personnel.

Access to records at the primary and secondary repositories must be restricted and monitored through the system's software with its required logon, security procedures, and audit trail. If remote access is provided via external software, the applications must enter through the same protective security software as that used for local access.

In this case, reports are paper-based. The operator/user may record each significant step of the computer-related operation and/or critical outputs. Paper records and associated electronic records are retained, based the requirements of the applicable predicate rule. In hybrid systems, one approach to the *signature/records linkage* is by referencing the file(s) as part of the report. The reference may include the name, and creation date and time of the files associated with the batch report. Modifications to any referenced electronic file will invalidate the approval of the associated batch report.

It is very important to ensure that having a signed paper-based record (e.g., a report) associated with one or more electronic records, the electronic file(s) associated with these records must not be deleted. The Part 11 regulation is very specific regarding this issue. Electronic files in hybrid systems must be maintained electronically. The interpretation of this requirement in the medical devices CGMP regulations (21 CFR 820) is different. Part 820 requires that the 'results' of acceptance activities are recorded, but are not necessarily raw data, and these 'results' must have audit trails. This interpretation is contained in the medical device quality system preamble (pp. 52631 and 52646).

Third Case: Implementation of Electronic Signatures

The application of an electronic signature refers to the act of affixing, by electronic means, a signature to an electronic record. Part 11 refers to two types of nonbiometrics-based electronic signatures: password/ID code-based signatures, and digital signatures. In addition to these methods, authentication systems (e.g., passwords, biometrics, physical feature authentication, behavioral actions, and token-based authentication) can be combined with cryptographic techniques to form an electronic signature.

Signed records must first conform to the requirements established in Part 11 Sub-Part B. In addition, verification and/or testing is required on the following:

1. Certification to FDA
2. Electronic signature security
3. Electronic signatures without biometric/behavioral identification
4. Electronic signatures with biometric/behavioral
5. Operational checks identification
6. Signature manifestation
7. Signature purpose

8. Signature binding and linkage
9. Certification to FDA

A *certification* must be sent to the FDA by the company, which informs the FDA that electronic signatures will be used. This certification can be a global statement of intent, meaning that a single company certification can cover all systems, all applications, all of the electronic signatures, for all employees in a company, for all the firm's locations anywhere in the world. The Regulatory Affairs Unit of the company will usually send this certification to the FDA.

Systems that use a password/ID code-based *nonbiometrics electronic signature* are common, and use both a PIN and a password as authentication and as an electronic signature. The combination of the PIN and password must be unique. Generally, companies and organizations ensure that the PIN is unique so that if, by coincidence, two persons create the same password, this does not result in two identical electronic signatures. Furthermore, firms generally establish unique (but not confidential) PINs.

Biometrics is the use of physical characteristics such as a voiceprint, fingerprint, hand geometry, and iris or retina pattern, as a means of uniquely identifying an individual. When biometrics measures are applied in combination with other controls (such as access cards or passwords), the reliability of authentication controls takes a giant step forward. The use of biometrics/behavioral identification is less common in the FDA-regulated industry. This access control technology does not change the authentication of users; it changes the implementation of how to authenticate users.

Part 11 requires a *signature manifestation* to be executed as part of printing and displaying signed records. The regulation requires that the name of the signer, the date and time of the executed signature, and the purpose associated with the signing, be displayed/printed in human readable forms.

As part of the action of signing an electronic record, the purpose of the signature must be identified and be an element of each signed record. In the drug CGMP arena, there are relatively few purposes for a signature and these include 'authored by,' 'reviewed by,' and 'approved by.' The execution of an action such as a production step does not require an electronic signature, and can be documented via the audit trail.

Special attention must be paid to the signature binding requirement of Part 11. The current requirement states that electronic signatures must not be replaced or removed. This requirement ensures the integrity of the signed records. In addition to the access control technologies and procedures, signature binding needs supporting tools for verification.

For example, PKI technology uses hashing algorithms and keys to demonstrate the integrity of signed records. The digital signature is linked to the electronic record by incorporating an instance of the record, into the signature itself. The link must be retained for as long as the record is kept, which may be long after the signer has departed from the company.

The records/signatures linking using password/ID code-based signatures is either based on the use of software locks, on storing the electronic signature in a database table separate from its associated record, or on storing the signature within the subject electronic record. All these schemes require procedural and technological controls to ensure that the requirements of Part 11.70 are met.

Criticality and Complexity Assessment[1]

PURPOSE

The Criticality and Complexity Assessment worksheet provides the guidance needed to determine whether or not validation is required for new or modified computer systems. If validation is required, the worksheet assigns it a level from 1 to 3. The validation level determines the amount of documentation, testing, resources, and training that may be required for the project.

There are five steps involved in completing the criticality and complexity assessment:

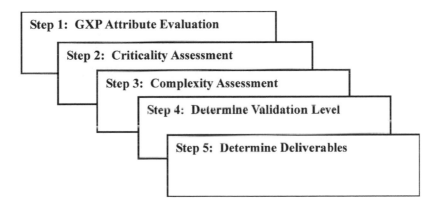

Step 1: GXP Attribute Evaluation

Step 2: Criticality Assessment

Step 3: Complexity Assessment

Step 4: Determine Validation Level

Step 5: Determine Deliverables

At the end of the assessment, the Deliverables Matrices detail the documentation that may be required for each level of validation.

An outline of the criticality and complexity assessment results must be included in the Requirements/Specification document for Validation Level 1, or the Validation Plan for Levels 2 and 3. In those cases where the Requirements/Specification document and Validation Plan are not required, an outline of the criticality and complexity assessment result must be included in the qualification protocols.

The selection of deliverables for each validation level may differ from those indicated in the Deliverables Matrices depending on the project documentation deliverables policies used by each regulated company.

---

[1] This Criticality and Complexity Assessment is taken from McNeil Consumer & Specialty Pharmaceuticals (MCSP) computer systems validation methodology. Naming of deliverables in this appendix is taken from MCSP methodology. Naming of deliverables in this book is not consistent with MCSP methodology.

VALIDATION MODEL DECISION TREE

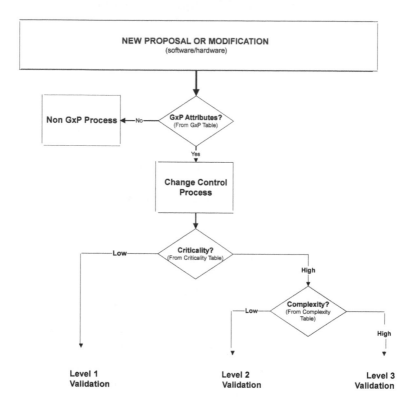

Step 1: GxP Attribute Evaluation

Is the system being used to perform a regulated operation, or to feed information to another system supporting regulated information activities? Does the computerized system impact or affect any of the following attributes? If not, then it is not within the scope of this assessment.

		YES or NO
1.1	Does the system control a manufacturing and/or testing process that has a direct impact on product quality? Product quality attributes include identity, efficacy, strength, dosage, safety, and purity.	
1.2	Does the system produce data that is used to release components or materials?	
1.3	Does the system create, modify, maintain, archive, retrieve, or transmit GxP-related information in electronic format that is required by regulatory agencies?	

- If the answer to *any* of the questions in Step 1 was 'yes,' then proceed to **Step 2**.
- If the answers to *all* of the questions in Step 1 were 'no,' then:
  a. The computer system does not perform functions in a regulated operation and the computer systems validation written procedures are not applicable.
  b. If preferred, the computer system may be subject to an alternative SLC, one that is not as rigorous as that required by the applicable regulation.
  c. Attach support documentation with the evaluation, if applicable, and store in the project file.

  Proceed to Step 2.

Step 2: Criticality Assessment

What would be the impact on the regulated area if the data in the system were corrupted, or if the system produced incorrect results? The higher the impact, the more extensive the validation needs to be.

Criticality Matrix: What is the degree of criticality of the system? Enter either Low or High in the Rating box as applicable.

		YES or NO
2.1	Has the system been implicated or found deficient by any regulatory authorities?	
2.2	Does the system (or should the system based upon GxPs) utilize any type of electronic signature?	
2.3	Can the system be classified as any of the following: Laboratory Data Systems Adverse Event Reporting System Supervisory Control and Data Acquisition System Stability Reporting/Tracking System Nonconformance/Investigation/Deviation System Complaint System Electronic Batch Record System Calibration/Preventative Maintenance Tracking System ERP/MRP Label Management System Facility Monitoring and/or Control System	
2.4	Does the system or modification/change create, modify, maintain, archive, retrieve or transmit data that has been predetermined to be either a critical parameter or data that is used to release a batch?	
2.5	Does the system or modification/change create, modify, maintain, archive, retrieve or transmit data used in regulatory submissions?	
2.6	Does the system or modification/change dictate the operation of a critical process?	
	Criticality	**Rating High/Low**
• •	If the answer is YES to any of the questions 2.1–2.6, then the group is considered to be a high-criticality group. If the answer is NO to all of the questions 2.1–2.6, then the group is considered to be a low-criticality group.	

Proceed to Step 3.

Step 3: Complexity Assessment

How complex is the system? Would the user immediately detect any problem? For example, if a system is producing three forms, and the user can immediately detect the wrong form, the extent of validation may be minimal. Conversely, if the system were performing complex calculations to determine if a product passes or fails product safety tests, and there is no simple way to check these calculations, the extent of validation would be greater.

Complexity Matrix: What is the degree of complexity of the system? Enter either Low or High in the Rating box, as appropriate.

		YES or NO
3.1	**Technology**: Does the system or system modification/change require the application of existing technology that is new to the company or that involves the use or development of a new technology that requires specialist skills and expertise?	
3.2	**Resources**: Does the system or system modification/change to be implemented require a dedicated team or a cross-functional team?	
3.3	**Software**: Does the system or system modification/change affect other subsystems? Is there a large volume of code, or are there very complicated interfaces associated with the system?	
3.4	**Infrastructure**: Does the system or system modification/change impact multiple or company-wide functions, or is new infrastructure required?	
	Complexity	**Rating High/Low**
• •	If the answer is YES to any of the questions 3.1–3.4, then the system is considered to be a high-complexity system. If the answer is NO to all of the questions 3.1–3.4, then the system is considered to be a low-complexity system.	

Proceed to Step 4.

Note: In the complexity matrix:

- **Technology** refers to *the level of technology (either process or equipment) required to develop, implement, or change a system.*
- **Resources** refer to *the people, equipment, and facilities/utilities required to develop, implement, or change a system.*
- **Software** refers to *the complexity of the Software Development effort required to implement or change a system.*
- **Infrastructure** refers to *the infrastructure (e.g., organizations, systems, and procedures) required to implement, operate, or change a system.*

Step 4: Determine the Required Level of Procedures

Enter the criticality and complexity rating determined in Step 2 and Step 3.

Criticality	Complexity

Using the table below, determine the validation level.

Criticality	Complexity	Validation Level
Low	N/A	LEVEL 1
	N/A	
High	Low	LEVEL 2
	High	LEVEL 3

Enter the Validation Level.

Level

Enter the Category of Software. Note: same as GAMP.

Category of Software

Proceed to the Deliverables Matrices that follow and select the matrix that is applicable for the Validation Level that has been determined. This will enable the documentation requirements to be determined. Then proceed to the page after the Deliverable Matrices and enter the documents required for the computer system.

Deliverables Matrices

The following three matrices detail the deliverables for each level of validation. There is one matrix per validation level.

Level 1: Deliverables Matrix

Software Categories (GAMP)

Phase	Documentation Requirements	1	2	3	4	5
Requirements	Qualification and CSV Plan Vendor Qualification Plan/ Certification User Requirements Functional Specifications			NR1* R1*	NR1 R1	NR1 R1
Design	System Design Specifications Traceability Analysis				R2	R2
Implementation	Unit Test Documentation & Report Code and Code Review Documentation In-process Vendor/System Developer Audit Report Integration Test Procedures					
Test	Integration Test Report IQ Protocol and Report OQ Protocol and Report PQ Protocol and Report Vendor Qualification Report Validation Report	R3	R3 R3	R3 R3	R3 R3	R3 R3

NR   = Required for new systems
Rn   = Required documents for group 'n' may be combined into a single
         document (i.e., one document for all R1 deliverables)
*   = A comprehensive user manual may be substituted for the user
         requirements and/or functional specification
Blank = Not required

LEVEL 2: DELIVERABLES MATRIX

Software Categories (GAMP)

Phase	Documentation Requirements	1	2	3	4	5
Requirements	Qualification and CSV Plan Vendor Qualification Plan/ Certification User Requirements Functional Specifications			R1 NR1* R1*	R1 NR1 R1	R1 R1 R1
Design	System Design Specifications Traceability Analysis				R2 	R2 R2
Implementation	Unit Test Documentation & Report Code and Code Review Documentation In-process Vendor/System Developer Audit Report Integration Test Procedures				NR3 R3	NR3 R3
Test	Integration Test Report IQ Protocol and Report OQ Protocol and Report PQ Protocol and Report Vendor Qualification Report Validation Report	 R4	 R4 R4	 R4 R4 R4 R5	 R4 R4 R4 R5	 R4 R4 R4 R5

NR   =   Required for new systems
Rn   =   Required documents for group 'n' may be combined into a single
              document (i.e., one document for all R1 deliverables)
*   =   A comprehensive user manual may be substituted for the user
              requirements and/or functional specification if approved in the
              qualification/validation plan.
Blank =   Not required

Level 3: Deliverables Matrix

Software Categories (GAMP)

Phase	Documentation Requirements	1	2	3	4	5
Requirements	Qualification and/or CSV Plan Vendor Qualification Certification (New/Modified Applications) User Requirements Functional Specifications			R1 NR3* R3*	R1 R2 NR3 R3	R1 R2 NR3 R3
Design	System Design Specifications Traceability Analysis				R4 R5	R4 R5
Implementation	Unit Test Documentation & Report Code and Code Review Documentation In-Process Vendor/System Developer Audit Report Integration Test Procedures				NR6 R6 R7	NR6 R6 R7
Test	Integration Test Report IQ Protocol and Report OQ Protocol and Report PQ Protocol and Report Vendor Qualification Report Validation Report	 R8	 R8 R8	 R8 R8 R8 R9 R10	 R8 R8 R8 R9 R10	 R8 R8 R8 R9 R10

NR = Required for new systems
Rn = Required documents for group 'n' may be combined into a single
   document (i.e., one document for all R1 deliverables)
*   = A comprehensive user manual may be substituted for the user
   requirements and/or functional specification if approved in the
   qualification/validation plan.
Blank = Not required

Deliverables Matrix:

Software Categories (GAMP)

Phase	Documentation Requirements	1	2	3	4	5
Requirements	Qualification and CSV Plan Vendor Qualification Plan/ Certification User Requirements Functional Specifications					
Design	System Design Specifications Test Plan Traceability Matrix					
Implementation	Unit Test Documentation & Report Code and Code Review Documentation In-Process Vendor/System Developer Audit Report Integration Test Procedures					
Test	IntegrationTest Report IQ Protocol and Report OQ Protocol and Report PQ Protocol and Report Vendor Qualification Report Validation Report					

Sample Development Activities Grouped by Project Periods

Using a critical and complex custom-built system as an example, and the conventional waterfall model (see Chapter 7, Figure 7–1) as the basic framework for specification, design, and testing development methodology, the following sections contain a detailed description of a development methodology grouped by periods. An overview of this appendix is provided in Figure E–3.

CONCEPTUALIZATION PERIOD

The system life cycle starts with the conceptualization period. This period is critical and is the highest level of design, and is primarily the responsibility of the system owner. The purpose of the conceptualization period is to establish an integrated model of the operation to be automated or re-engineered. This model provides the criterion to company senior management, enabling them to make a business and/or regulatory decision on whether to continue with a project.

The user's needs are analyzed in order to determine appropriate prospective solutions. The prospective solutions are high-level solutions identifying major constraints, system interfaces, user interfaces, and proposed technologies. These solutions are also used to assess the criticality and complexity of each proposed system component (refer to Appendix D for more information). This assessment will provide information on the resources and documentation requirements for the project.

During the conceptualization period, the scope of the system is defined based on the strategic objectives contained in a marketing proposal or similar document.

Supporting information is obtained for this period from interviews and by reviewing documentation generated by operations, R&D, QA, and Engineering/IT. Part 11 should be included as part of the integrated model of the operation to be automated or re-engineered.

During the operational life of the computer system, a change request to enhance or amend the computer system will result in this period being initiated again and the original concept being reviewed and updated if necessary.

The outputs of the conceptualization period can include operational checks/sequencing models. A new or updated concept document is produced during this phase.

DEVELOPMENT PERIOD

The purpose of the Development Period is to define a structured process to specify, design, build, test, install, and support the business operation to be automated or updated.

The output of the development period may include the following:

- plans (e.g., project, validation)
- written and approved requirements (e.g., user's, functional, design) that describe what the application is intended to do and how it is intended that it should do it

- test results and an evaluation of how these results demonstrate that the predetermined requirements (e.g., user's, functional, design) have been met
- system acceptance for use

Requirements Gathering and Specification Development

The scope of the system generates many of the support requirements for the operations, usually provided by the system owner. The term 'requirement' defines a bounded characterization of the scope of the system, and contains information essential to support the operation/operators. Some of these requirements include functional capacity, execution capability, operational usability, information needed to support validation, installation, commissioning, SLC documentation required, user's manuals, training, maintenance manual, system maintenance, system test plan, acceptance criteria, and regulatory compliance (e.g., Part 11).

The majority of computer system developments fail due to poor gathering of requirements. This situation negatively affects subsequent development activities and associated working products.

The system requirements are captured in the requirements specification deliverable, which describes what the system is supposed to do from (a) the process, (b) the user's, and (c) the compliance perspectives. The requirements specification deliverable may be used as a framework to select the supplier/ integrator and to develop the PQ protocol.

The requirements specification deliverable must include an overview of the process in order to familiarize the application software developer with the user, process and data acquisition requirements of the system, as well as special considerations for the project. The system functionality must be well defined at the outset, in order to provide the prospective supplier/ integrator with enough information to provide a detailed and meaningful quotation.

Specifically, on data acquisition systems, the requirements specification deliverable must include definitions of (i) the data to be collected, (ii) how the data will be used, (iii) how it will be stored and retention requirements, (iv) data security requirements, and (v) where each operation will be completed. The requirements specification deliverable addresses the following:

- the scope of the system and strategic objectives
- the problem to be solved
- process overview, sequencing requirements, operational checks
- sufficient information to enable the supplier/integrator to work on a solution to the problem (e.g., device-driven sequencing, the methods required of the presentation of data, data security, data backup, data and status reporting and trending, etc.)
- redundancy and error-detection protocol
- operating environmental
- interfaces (e.g., to field devices, data acquisition systems, reports, and HMI), I/O lists, communications protocols, and data link requirements
- information gained from operators and supervisors on the system design requirements and expectations in order to influence how the system is designed and operated
- type of control and process operations to be performed
- data storage requirements
- transaction/data timing requirements and considerations
- regulatory requirements (e.g., validation, Part 11)
- preliminary evaluation of the technology

- feasibility study and preliminary risk assessment
- safety and security considerations
- Part 11, security, other requirements
- nonfunctional requirements (e.g., SLC development standards, programming language standards, program naming convention standards, etc.)

Each requirement in the requirements specification deliverable must have the following attributes:

Unambiguous	All requirements must have only one interpretation
Verifiable	A human being or a machine must be able to verify that the system correctly implements the stated requirements
Traceable	The requirements must be traceable from other design and test documents
Modifiable	Unanticipated changes must be able to be made easily
Usable	The URS must be usable by not only the development team but also subsequent maintenance teams who will be called upon to modify and change the system
Consistent	Individual requirements must not conflict with each other
Complete	It must contain clear descriptions of all features and functions of the system. It must clearly contain definitions of all known situations the system could encounter

The requirements specification deliverable is tested in an operating environment during the PQ and must include the verification of the procedural controls associated with the system, as identified in the requirements specification deliverable.

The following section provides an example of the general content of a requirements specification deliverable and the type of information to be included.

- **Functions**
  This subsection defines the required system functions, modes of operation, and behaviour. It should address the following:
  - functions required. Information on the process or existing manual system should be included here, if not covered adequately elsewhere
  - calculations, including all critical algorithms
  - modes of operation (e.g., start-up, shutdown, test, fallback manual, automatic)
  - performance and timing requirements. These should be quantitative and unambiguous
  - failure response. The action required in case of selected software or hardware failure
  - safety and security

- **Data**

  This subsection states the data-handling requirements. It should address the following:
  - definition of the data, including identification of critical parameters, valid data ranges and limits, and how out-of-range data should be handled
  - capacity requirements
  - format requirements (e.g., when transferring data between systems and for meeting Part 11 data inspection requirements)
  - access speed requirements
  - archive requirements

- **Interfaces**

  This subsection defines any system interfaces and contains the following subsections:
  - interfaces with users. These should be defined in terms of roles (e.g., plant operator, warehouse administrator, manager, etc.) and/or functions as appropriate
  - interfaces with other systems
  - interfaces with equipment (e.g., sensors/actuators). This includes I/O listings for process control systems

- **Environment**

  This subsection defines the environment in which the system is to work. It contains the following subsections:
  - layout. The physical layout of the plant or work place may have an impact on the system (e.g., long-distance links, space limitations, proximity to electrical noise sources, etc.)
  - environmental conditions (e.g., temperature, humidity, dusty or sterile environment, etc.)

- **Constraints**

  This section defines the constraints on the specification of the system. It contains the following subsections:
  - timescales and milestones, as appropriate
  - compatibility. This takes into account any existing systems or hardware, and any company or plant strategies
  - availability. This states the reliability requirements, and the maximum allowable periods for maintenance or other downtime
  - procedural constraints (e.g., statutory obligations, legal issues, working methods and user skill levels)
  - maintenance (e.g., ease of maintenance, expansion capability, likely enhancements, expected lifetime, long-term support)

- **Life cycle**

  This section defines any requirements concerning the development life cycle. It contains the following subsections:
  - development (e.g., procedures for project management and quality assurance, mandatory design methods, design review requirements, etc.)
  - testing (e.g., special testing requirements, test data, load testing, simulation, reporting of the test results, etc.)

- **Deliverables**

  This section defines what deliverables are required for the project. It should address the following:

- how deliverable items are to be identified
- in what form deliverables are to be supplied (e.g., format and media)
- documents. What the supplier is expected to deliver (e.g., functional specification, testing specifications, design specifications, design reviews, test results, etc.)
- data to be prepared or converted
- tools
- training courses
- archiving facilities

- **Support**
  - This section defines what support is required after acceptance

- **Glossary**
  This section contains definitions of terms, which may be unfamiliar to the reader of the requirements specification deliverable

- **Appendices**
  Any additional information that may help the computer system supplier/developer to provide a quotation can be provided as appendices

Develop a Validation Plan

The validation plan is discussed in Chapter 8.

Conduct a System (Hardware and/or Software) Risk Analysis

Software development is one of the most risk-prone management challenges. Risk factors can negatively impact the development process and, if ignored, can lead to project failure. To counteract these factors, system risk must be actively assessed, controlled, and reduced on a routine basis.[1]

Risk is defined as 'the probability of an undesirable event occurring and the impact of that event if it does occur.' The result of this analysis will influence the degree to which the system development, implementation, and maintenance activities are performed and documented. By evaluating the system risk analysis, the system owner may uncover potential problems, which can be avoided during the development process. The chances of a successful, if not perfect, system implementation are improved.

There are seven steps necessary for assessing and managing risks:

- a description of the risk
- the cause(s) of the risk
- the level of concern regarding the risk based on a qualitative estimate, including any definition of terms used
- the likelihood of occurrence of the risk based on a qualitative estimate, including the definition of terms used

---

[1] *Guidelines for Successful Acquisition and Management of Software Intensive Systems*, Vol. 1, Software Technology Support Center, DAF, February 1995.

- the method(s) of control used to eliminate or mitigate the risk, e.g., a change in the design specification, alarm warning and/or error messages, implementation of a manual process and/or workaround
- the traceability of each method of control to the corresponding critical requirement specified in the requirements specification deliverable
- monitor and manage the implementation of the requirements identified during the risk analysis

Some of the items to consider when conducting the risk analysis include:

- regulatory areas of the system
- size of the computer system and number of users
- complexity of the system in terms of the design
- type of data the system handles
- functionality of the system
- interaction with and impact on existing systems
- vendor reliability
- effect/impact of system noncompliance
- safety to the customer, employee, and community
- cost versus benefit of the system

The results of the risk analysis review must be documented. To adequately analyze the risks, a comprehensive requirements specification deliverable is required. The best time to perform an initial risk analysis is immediately after the project team has completed the review of the requirements specification deliverable. The review of the solutions agreed upon and implemented as a result of the risk analysis should be performed during a technical design review.

During the design and development of computer systems or application software to be used as components in medical devices, to be manufactured and/or sold in the USA, consideration must be given to meeting the requirements for design in 21 CFR Part 820 of the FDA regulations.

Conduct a Supplier Audit of the Computer Technology Suppliers and/or Developer

Refer to Chapter 17.

Develop a Computer System Specification

The computer system specification includes information on how the operation can be controlled using the features in the selected application package. A review of the process must be conducted to familiarize the computer technology supplier and/or developer (external or in-house) with the requirements defined in the requirements specification deliverable. This system view is essential, particularly when the software must interface with other elements such as hardware, people, and databases. The review of the process includes the further gathering of requirements at the system level, and top-level design.

The computer system specification deliverable for the system identifies the detailed requirements to be incorporated into the design and tested during the FAT, SAT, and/or OQ. The computer system specification takes the information provided by the requirements specification deliverable, and provides the following:

- functions defined in detail
- interfaces further described
- design constraints and performance characteristics established
- database characteristics identified
- test and validation criteria defined

The documentation generated during this phase may include data flow diagrams, data definitions, detailed descriptions of the processes and functions, performance requirements, and security considerations.

The elements of the computer system specification deliverable relate directly to the documentation and test cases contained in the OQ protocol. Once the computer system specification deliverable is approved, it is possible to start the detailed design phase and to generate the FAT, the SAT, and/or the OQ documentation.

The following is an example of the general content of the computer system specification deliverable for a system. If no requirement has been specified in the requirements specification deliverable for a particular section, the section must still be included but must state 'Not Applicable.'

- **Introduction section**
  This section contains the following information:
  - who developed the computer system specification deliverable, under what authority, and for what purpose
  - relationship to other documents

- **Overview section**
  This section states the essential system functions, and interfaces to the outside world. It should cover the following:
  - key objectives and benefits
  - high-level description. This should provide breakdown of the computer system into the main subsystems (hardware and software)
  - the main interfaces between the system and the outside world
  - assumptions. This should state any design or implementation assumptions (e.g., standard packages, operating system, hardware)
  - nonconformance with the requirements specification deliverable. Any divergence between the computer system specification deliverable and the requirements specification deliverable must be clearly noted

- **Functions section**
  This section takes the high-level description, and breaks it down to the level of the individual functions. It describes the functions and facilities to be provided, including specific modes of operation. This description is based on the features implemented in the selected application. The following aspects should be addressed:
  - the objective of the function or facility, and the details of its use, including the interface with other parts of the system. Critical calculations and algorithms should be highlighted
  - the performance (e.g., response, accuracy, and throughput), which should be quantitative and unambiguous
  - safety and security. The topics covered may include: the action to be taken in case of selected software or hardware failures, self-checking, input validation, redundancy, access restrictions, time-outs and data recovery

- functions that are configurable and any limits within which the configuration can take place
- clause level traceability to specific requirements in the requirements specification deliverable

- **Data section**
  This section defines the data on which the system is expected to work. The following aspects should be addressed:
  - definition. The data should be defined in a hierarchical manner with complex objects being built up from simpler objects (e.g., files of records; complex data types defined in terms of simple data types). Critical parameters should be highlighted
  - access (e.g., which subsystems need read or write access to each data item, access method, speed and update time, read/write interlocks)
  - data capacity, retention times, and details of how the data should be archived

- **Interfaces section**
  This section defines any system interfaces. It contains the following subsections:
  - interface with users defined in terms of roles (e.g., operator, administrator, clerk, and system manager). Topics to consider include: the facilities available, suitable types of peripherals, the general format of displays and reports
  - interface with other systems. Topics to consider include: data transmitted and received, rate of transfer, data protocol, and the security of the interface
  - interface with equipment (e.g., sensors, plant equipment). Topics to consider include: data transmitted/received, data format, validation, and error checking

- **Nonfunctional attributes section**
  This section defines how the system will meet nonfunctional requirements. It contains the following subsections:
  - availability (e.g., reliability, redundancy, error checking, standby operation)
  - maintainability (e.g., expansion and enhancement possibilities, spare capacity, likely changes in environment, system life expectancy)

- **Glossary section**
  This section contains definitions of terms, some of which may be unfamiliar to the readership of the computer system specification deliverable.

- **Appendices**
  Where appropriate (e.g., small systems), appendices may be provided to define hardware and software specifications.

Develop a Technical Design Specification

The design process and resulting deliverable are based on, and clause traceable to, the information in the requirements specification deliverable. The technical design specification must specify how the combined hardware and software products will satisfy the requirements.

Also specified in the technical design specification deliverable are the subsystems, components and interfaces, data structure, design constraints, algorithms and system decomposition of the computer system to be supplied by the project. The objective of a good modular design or partitioned design is to separate both data and functions into modules. Each module is called a unit. This concept is very important when linking the designed units and when testing these units. Modular design is also a key concept when developing computer systems using an object-oriented development methodology.

The technical design specification deliverable for the computer system is a document defining the system in sufficient detail to enable it to be built, and to enable code to be produced for it, including:

- the system architecture and functions
- hardware/software installation specifications and configuration
- components
- modules
- interfaces (machine and human)
- drawings or pictorial representation of the hardware that associates part of the system and its interfaces to other hardware
- system operation under both normal and abnormal conditions
- logical and physical structure of the software programs
- safety and security
- design and programming standards

The technical design specification deliverable may include block diagrams, flow diagrams, state transition diagrams, loop control calculations, and narrative descriptions.

Depending on the technology available and the cost, automated functions can be assigned to the computer hardware or software. The computer hardware can be further decomposed into a number of subelements.

The software design concepts such as 'modular,' 'top-down,' and 'decomposition' of the design requirements have been used in the computer science fields for over 20 years. However, the application of these concepts can vary.

The support and maintenance of the software is simplified when consistent formats and techniques are utilized. Troubleshooting programming logic not implemented in a structured format is frustrating and tedious.

Developing a system that provides for ease of operation and maintenance requires that the software engineers, user owners, and end-users agree on the:

- design of the operator interface including overview displays
- format of the program documentation and operational manuals
- partitioning of each program into simple modules that can be reused throughout the overall application software

It is imperative that an agreement is reached on a structured approach before any programming/configuration of the software begins. This approach should provide the basis for a design specification deliverable which defines the system development scope.

Typically, 80% of the programming or configuration process can be partitioned into modules that are reusable throughout the system. Once a module has been developed and tested using simulation techniques, it can be reused with little risk. It is important that these modules are well documented, so that persons wishing to make use of the module can easily understand them. It is important that the misapplication of modules is minimized.

The remaining 20% of the programming/configuration process may represent custom logic unique for that particular area of the application. Custom logic can be made easy to understand

and maintained through the use of consistent formats, variable utilization, and effective documentation within the logic.

When a structured and modular architecture is used, the software becomes easier to read, troubleshoot, modify, and extend. It quickly becomes obvious that a structured approach is vital in these flexible yet complex systems.

The logical structures or implementation contained in the technical design specification deliverable must be reviewed during a design review. A design review is performed based on an approved procedure. The objective of the design review is to verify that the logic structure including the modularity, reuse, functionality, and maintainability of the implementation, and the clause traceability of the computer system specification deliverable to the technical design specification deliverable. The technical design specification deliverable also contains information on how the risks inherent in the system design were resolved.

The system design is released for implementation once the technical design specification deliverable has been approved. All changes made to the computerized system design after the approval of the technical design specification deliverable are reviewed and documented using a change management system/control.

During each software or system development step, the developer provides documented evidence that implementing the requirements specified in the requirements specification deliverable developed the product. During the design, the in-process (internal) audit must be carried out in order to verify that the design of the computer system satisfies the requirements described in the computer system specification deliverable, and that the code has been developed in accordance with the technical design specification deliverable.

A comprehensive design review or design qualification should be performed to evaluate the design capabilities to fulfill the requirements, and a business case design solution prepared.

The physical implementation of the system described in the technical design specification deliverable must be verified during the IQ.

Design key practices can be found in Appendix I.

Consideration of the Human Factors in the Design of Operator Interfaces

Poor operator interface design induces errors and inefficiency among even the best-trained operators, especially under conditions of stress, time constraints, and/or fatigue. Although labeling (e.g., user documentation) is extremely important for good performance, even well-written instructions are cumbersome to use in conjunction with actual operation. In addition, it is difficult to write coherent documentation that describes awkward operating procedures.

Although both hardware and software design influence the operator's performance, the logical and informational characteristics provided via software are increasingly crucial. Data presented in an ambiguous, difficult-to-read, or nonintuitive manner poses the threat of an incorrect reading, misinterpretation, and/or improper data entry. An example of this could be a crowded display with cryptic identifiers, combined with a time lag between the operator response and the displayed responses. Such design characteristics overtax the user's abilities (e.g., memory, visual perception, decision-making, etc.), and resultant errors may have serious consequences.

Software Development

This is the lowest level of requirements decomposition. The technical design specification deliverable is translated into software source code. Source code is the human readable version of the list of instructions (programs) that enable a computer to perform a task. Source code is written in the programming language appropriate for the application to be generated.

Except for a very small number of programming languages, source code cannot be understood and executed by a computer system. Consequently, the source code is converted into a machine-executable form prior to being run. Machine-executable code is also known as an *object program*. This conversion into machine code is called translation, performed by another program called an assembler, compiler, or interpreter, depending on the programming language involved.

All these activities are directed by procedures containing programming standards, naming conventions, and the configuration settings for the compilers, assembler, and/or interpreters. In addition to the object program, the compilers/assembler used will list the results of the translation process, highlight any errors that may have been encountered, with possible resolution. An interpreter does not create any files or lists.

In a mature software development environment, software is developed as individual components or small units. Each unit is tested individually and integrated with other units to form a module. The accurate implementation of each module is achieved by following the technical design specification deliverable. The output of the software development process is an assembled system.

The software development process involves coding, debugging, and formal testing (code audits, unit testing, and integration testing) by the developer. The developer must comply with the established programming standards, where applicable, to ensure that the program has a uniform content and to assist with the ongoing understanding of the source code.

- **Code reviews, inspections, and audits**

  During and after the coding process, code reviews, inspections, and audits on application software and/or configurable products must be performed. Refer to Chapter 9. The objectives of code reviews, inspections, and audits are:
  - the verification of the absence of dead code (CPG 7132a.15)
  - confirmation of a modular design
  - verification of the conformance with the programming approach selected
  - verification of the design modification that were implemented as a result of the risk analysis
  - the features incorporated into the code in order for it to be efficiently maintained during its operational life

  Code reviews, inspections, and audits may make use of high-level system data flow diagrams (e.g., transaction flow diagrams) in order to determine the scope and depth of the reviews, inspections, and audits.

- **Conduct software module (unit) testing and integration testing**

  Figure E–1 depicts the essential activities of unit, integration, and system level (FAT, SAT, OQ, PQ) testing, and the relationships between them. These activities are discussed in the following sections.

  A unit test is a white box-oriented test, conducted on modules that have been structurally tested. This type of test involves the standalone software unit before it is integrated into a complete system or subsystem. The test cases are designed to look for errors in the logic design, data processing, and the statement of execution. Uncovered errors are removed, and the test case that located the error, and any other associated test case, re-executed.

  It is not always feasible to test a program with real equipment or hardware. In these cases, a program can be put under the required testing conditions and tested using input simulation.

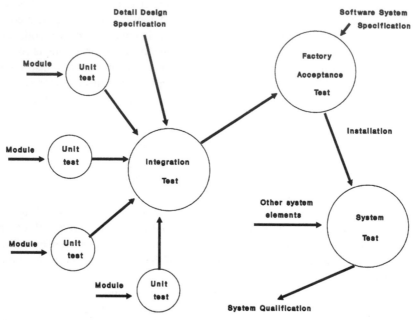

Figure E–1. The relationship between the testing activities.

The items that can be checked during unit test are as follows:
- that it functions as specified in the appropriate design document
- maximum or minimum limits testing
- reaction to incorrect inputs
- exception handling
- algorithm assessment
- adherence to documented standards

The software integration testing is an orderly progression from unit testing and incremental testing in which software, hardware elements, or both are combined and tested, until the entire computer system has been assembled.

The objective of the integration test is to uncover errors associated with software interfaces and/or the integration of the software with the hardware. Unit-tested modules are built into a program structure that has been dictated by the design documentation.

The characteristics to be evaluated during the software integration testing are:

• interface integrity
• functional validity
• information content

The software test strategy and detailed test scripts (protocol) are reviewed and approved by the Project Team. This review will verify that the test cases will stress the program interfaces, challenge the data boundaries and limits, and verify that the test cases are traceable to the appropriate clauses in the requirements specification deliverable. The review also verifies that the test cases adequately cover the risks identified in the Risk Analysis. The unit/ integration test results are documented by the programmer, and reviewed, approved, signed, and dated by QA.

Preparation of the Qualification Protocols

Qualifications are formal examinations of the completed system in order to determine and record the acceptability of the design, fabrication/development, and integration of the hardware/ software and associated utilities.

Qualification protocols are written documents prepared before conducting the qualification, describing the features of a particular application or item and how it should be tested. Qualification protocols identify the objectives, methods, and acceptance criteria for each test function contained in the applicable specification deliverable; and identifies who is responsible for conducting the tests. In addition, the protocols should also specify how the data is to be collected, reported, and analyzed to determine if the acceptance criteria were met. The protocol should be reviewed by personnel with an appropriate understanding of computer systems and the functionality of the indicated system. Following the review, qualified personnel must approve each protocol in accordance with the company quality assurance procedures. In the software engineering world, qualification protocols are equivalent to test procedures. Traditionally the scope of the qualification protocols encompasses the following:

- installation
- operation
- performance

The focus of qualification of focus protocols is always prospective and the detail contained in each dependent upon the validation tasks performed.

The contents of the FAT, SAT, and OO tests are similar. The differences between these activities are the test environment and the formality of the testing. Figure E–2 depicts an example of a qualification timeline that contains these activities. FAT, SAT, and OQ are described in Chapter 10. The successful execution of the protocols provides documented evidence that the computer system is performing as specified in the requirements specification deliverable, computer system specification deliverable, and the technical design specification deliverable.

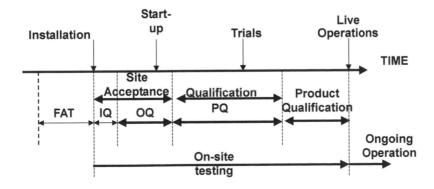

Figure E–2. Qualification Timeline.

The recommended contents of a protocol are listed below. The presence or exclusion of a particular item depends on the scope of the computer system validation.

a. Purpose or objective of the protocol
b. Scope of the validation activities
c. Definition of the system to be validated
d. Responsibilities for:
   - Creation of the protocol
   - Review/approval of the protocol
   - Execution of protocol
   - Review of the validation effort required
   - Approval/acceptance of the validation effort
   - The meaning of each signature
e. Validation rationale (including references to any company policies):
   - Discussion of the regulatory nature of the computer system
   - Reference to the company validation policies, methods, or standards
   - Strategy for the validation (e.g., phased)
f. Exclusions and assumptions
g. High-level description of the system:
   - Hardware components (the computer platform and any equipment that interfaces to or is controlled by the system)
   - Software components:
     – System software (operating system and utilities)
     – Noncustom-supplied software
     – Custom software
     – Network/communications software.
   - Interfaces to other systems or software
   - High-level diagram of the system hardware configuration (if applicable)
   - High-level diagram of the system functions (if applicable)
h. Prequalification documentation verification:
   - Statement of work or project initiation
   - Business, system, and functional requirements specification deliverable
   - Detailed system design specification deliverables (computer system and technical design specifications)
   - Development testing specification deliverables and associated results:
     – Structure/code evaluation (e.g., code review records)
     – Unit testing
     – Integration testing
     – System testing
   - Reference to any project or development standards and methods
i. Supplier qualification (if applicable)
j. System physical location/environment and requirements.
k. Hardware IQ:
   - Operating environment and parameters
   - Hardware components
   - Interfaced equipment (if applicable)
   - Network/communications hardware (if applicable)
l. Software IQ:
   - Operating system and utilities software
   - Noncustom-supplied software
   - Custom software
   - Network/communications software (if applicable)

m. OQ
n. PQ
  - Test plan (if size of project warrants a separate document of this type)
  - Verification of traceability to user requirements
  - Criteria and procedure for the identification and resolution of nonconformances that occur during testing
o. Security (physical, platform, and application)
p. Training
q. Operational procedures:
  - System usage
  - System support, maintenance, and monitoring
r. Development of a Requirements Traceability Matrix
s. Change and configuration management
t. Backup and restoration
u. Disaster recovery and alternative procedures
v. Incident reporting for the computer system (hardware and software) once it has been used in a production environment
w. Periodic review
x. Acceptance criteria for the successful completion of the system validation
y. Addenda, supplements, or revisions to the protocol
z. Validation report
aa. Review and approval of the protocol for execution

Since the protocol is prospective in nature, sections h through v contain the following information in each section:

- Identification of the function being tested or a unique test identification number or indicator
- Any prerequisites necessary to perform the test
- If using an electronic test tool, indicate the test script, electronic test file name, or other identifier
- The test script, which includes:
  - Steps or actions to be performed
  - The input data for the appropriate steps
  - The expected result or expected system response to the input
  - An area for entry of the actual test result or for entry of directive information that references an attached output (for example, an attached report, screen print, audit log, output from an automated test tool, etc.)
- Areas to indicate pass or fail for critical steps and initials of the tester
- The acceptance criteria for the entire test script
- Area for the date of the test
- Area for signature of tester
- Area for indication of pass or fail of the test script
- Area for nonconformance comments
- Area for the date reviewed
- Area for the signature of the reviewer
- Area for comments regarding any nonconformance resolutions

Hardware Integration and Program Build

Hardware integration involves the connection of hardware and network components based on the system hardware configuration. This configuration is specified in the computer system specification deliverable and further defined in the technical design specification deliverable. If the hardware integration is performed at the supplier's/integrator's site before the FAT, there is no need for a formal document to record the hardware installation. If the hardware integration is performed at the target or production environment, then formal documentation is required. In the context of FDA regulated industries, the formal document is called a hardware IQ.

Conduct a Factory Acceptance Test

The FAT, Figure E–2, is a mutually agreed acceptance test of the system between the supplier/developer and the user. It is performed by the computer technology supplier/developer at the supplier's/developer's site. The user will often witness this test to accept the system for delivery and to mark a contractual milestone.

The SAT has traditionally been seen as engineering documents and activities in support of the commercial and technical contracts, used as risk-reduction steps as opposed to qualification activities. As computer technologies suppliers and contract developers become better at working in a controlled environment, the complexity and coverage of the acceptance testing is improving. This is reflected in the current GAMP guide (GAMP4 2001), where qualification activities are shown as being completed during acceptance testing activities. As referenced in Sidebar 10–1, the FAT has the possibly of being OQ1, often a positive way forward as it reduces the amount of rework associated with testing/qualification activities. Obviously there are going to be some drawbacks using this approach, and is a balance to be evaluated thoroughly before starting any project or project phase.

The plus points include:

- qualification starts early and some elements can be completed in the supplier's/developer's facility. This could save on rework costs
- reduced effort of retesting
- modular testing with the appropriate level of rigor is possible

For example a system can be taken through all its exception handling routines in a simulated environment; this might be less practical on a 'live' system.

The minus points include:

- acceptance testing has to fall under the appropriate level of change control to reflect that it is a qualification activity
- testing and qualification planning is more complex, as the full system coverage has to be demonstrated
- content and documentation standards of the acceptance testing documents need to be capable of supporting qualification and need to be approved by QA

The chosen route should be planned accordingly and included as part of the Validation Plan. Typically, the FAT addresses the following:

- interfaces
- system functionality (e.g., sequences, limits, alarms, boundaries)
- system operability and performance
- initial user training, if applicable under the contract
- critical parameters
- Part 11 technical controls
  - electronic records management
  - logical security
  - management and security of user identification code and password (authentication and/or electronic signature)
  - generation of accurate and complete copies of electronic records
  - system date and time integrity and security
  - audit trails, metadata, display and reporting of audit trails
  - operational, authority and device checks
  - signatures management (nonbiometrics/biometrics)
  - signatures manifestation
  - multisigning
  - signatures binding
  - signatures sequencing

The FAT provides evidence that the hardware and software are fully integrated, that they operate as indicated in the computer system specification deliverable, and meet the expectations of the user as defined in the requirements specification deliverable. This final formal integration test should be completed in an environment very similar to the operational environment. The system can be subjected to a real-world environment by using emulators and/or simulators which mimic system interfaces. The user's representative should evaluate the supporting documents, the operation, system functionality, and system reliability.

If the FAT results and audits conform to the contractual agreements the system is accepted for shipment and installation. In particular, all identified critical problems and deviations found during the execution of the FAT must be corrected before shipment.

CONDUCT A SITE ACCEPTANCE TEST

The SAT is an acceptance test comprising system installation, start-up, operational testing, and handover to customer, performed by the supplier/ integrator in the operational environment. As with the FAT, the user's representative should witness the tests and evaluate the results of the SAT.

The testing carried out in the development environment is not sufficient to verify the integrity and performance of the system. The application software must also be tested under actual conditions of intended use in the target or production environment. If the application is to be supplied by a third party in accordance with the end-user's requirements specification deliverable, it should be possible to carry out a completely prospective validation, as would be the case for a system developed in house, providing the requirement for validation is identified at the beginning. As with all systems, the responsibility for validation lies with the system's owner.

Early agreement should be reached between customer/owner and supplier with regard to the documentation necessary to establish the validity of system parts under supplier control. In addition, the supplier should agree to inspection(s) by the owner, or if this would compromise the supplier's commercial interests, by an independent expert, employed by the owner.

During the SAT, many of the FAT verifications are re-executed. The only difference between the FAT and the SAT is that the FAT is performed under nonoperational conditions, whereas the SAT is performed under operational conditions. Depending on the configuration management controls established during the FAT, the field independent FAT verifications and associated test case results may not be repeated during the SAT.

One key test in the SAT is the configuration and qualification of the hardware.

Depending on the type of computer system involved, this is possibly the first time that the complete system hardware has been assembled. The hardware configuration provides an equipment installation, configuration, and inventory list for each device and/or item of equipment associated with the computer system and network hardware. The supplier's manuals for each piece of hardware must be provided. The hardware qualification includes:

- system description and schematic drawings
- instruments list/schedules and drawings
- computer system hardware list (including part/serial numbers), purchase orders, and requirements/data sheets
- I/O lists/schedule
- wiring checks
- loop diagrams, cable schedules and routing diagrams, panel and junction box/termination layouts
- 'full' calibrations certificates that list the test equipment and its traceability to national standards, and include the limits of uncertainty results
- verification of hardware integration
- security verification
- execution of system diagnostics
- control equipment configuration verification
- security verification (physical)
- control equipment interface verification

During the software IQ the developer will produce a plan for installing the software in the target environment as specified by the technical requirements specification deliverable. The resources and information necessary to install the software product must be determined and be made available. The developer must assist the system owner and maintenance personnel with the set-up activities: where the installed software product is replacing an existing system, the developer must support any parallel operation activities required by the contract. The installation plan and procedures must be documented.

The developer must install the software product in accordance with the installation plan and procedures. It must be ensured that the software code and databases initialize, execute, and terminate as specified in the technical design specification deliverable. The installation events and results must be documented.

The software IQ includes:

- man–machine-interface installation verification
- database and configuration verification
- version verification for all software used by the computer system
- a list of the individual manuals that are applicable to the computerized system
- a list of all computer software for specific processor modules associated with the unit; this will include the source code, as applicable, and samples of the graphic displays for various situations

- workstation system configurations
- a text copy of the latest revision of each application program
- a copy of the module configuration database form for each module configuration type
- a computer printout of the final field device point configurations

The software OQ includes all of the factory acceptance testing and, specifically, the following:

- A fallback plan must be devised for recovery of existing services in the event that introduction of a new system or new functionality to an existing system causes service degradation or interruption
- If database conversion, migration, or preloading with data has to occur prior to OQ testing, the verification of these activities and their associated data may be addressed in either a pre-OQ test or during the OQ. Any data migration tools should be written under the control of a software quality management system. The data migration process should be verified and the verification performed before final data loading to the new system
- The Part 11 technical controls tested at the FAT
- All major functions and interfaces must be exercised
- Catastrophic recovery verification
- Worst-case conditions for the application software (*Note:* As defined by the FDA the worst case is '*a set of conditions encompassing upper and lower processing limits and circumstances, including those within SOPs, which pose the greatest chance of process or product failure when compared to ideal conditions.*')
- Man–machine-interface functional verification
  - Boundary values
  - Invalid values
  - Special values
  - Decision point and branch conditions
- Tests on the generation of reports
- Hardware redundancy verification
- Data acquisition and database access controls
- Trending, alarm and event processing
- Interface with other applications
- Timing (if applicable)
- Verification of security protection mechanisms (Logical)

In situations where the SAT is conducted according to approved protocols, the test results are documented properly and reviewed, approved by the user and QA, it may not be necessary to repeat the SAT tests during the IQ and OQ. The IQ and OQ should still identify all of the necessary tests, but result areas can simply reference the appropriate test carried out during the SAT, providing evidence of successful completion.

The IQ must identify the hardware, software, and interfaces used by the computer system and demonstrate their installation, in accordance with the design drawings, specifications, and supplier recommendations. These requirements are established and documented in the technical design specification deliverable.

The OQ must demonstrate that the functionality of the system is as established in the computer system specification deliverable. During the OQ, all systems interfacing with the computer system must be operational throughout the entire process. Document evidence must be obtained verifying execution of the various steps and functions detailed in test. The protocol must be approved and completed by the appropriate personnel.

Conduct a Performance Qualification

The computer system needs a comprehensive PQ test designed to determine the degree of accuracy with which the system conforms to the requirements specification deliverable. The computer system PQ should not be confused with the process PQ and the product qualification as described in the FDA Guideline on Principles of Process Validation. The PQ occurs under operational conditions but not as part of the actual operational process.

The specific test cases include:

- Verification of the system documentation, all procedural controls, migration, and production implementations. Procedural controls include Part 11 controls not covered by the technical controls
- Determination of the accuracy of the system in receiving, recording, storing, and processing the electronic records
- Determination of the system accuracy in arriving at the appropriate disposition decision based upon the data received by the system
- Determination of the integration of the computer system into the operating environment (refer to Chapter 2, Figure 2–1), including technologies, procedural controls, controlled processes/operations, people, and documentation
- Operational checks relating to electronic signatures, if applicable

The PQ must cover the full operating cycle of the computer system and operating process and be performed following an approved protocol.

Prepare Qualification(s) and Project Report(s)

Once the protocols have been completed, the test results and data need to be formally evaluated. The written evaluation needs to be presented clearly, in a manner that can be readily understood. The report should also address any nonconformances or deviations to the validation plan encountered during the qualification and their resolutions. The format of the report should be similar to the structure of the associated protocols. The qualification testing should be linked with the acceptance criteria in the relevant specification deliverable, so that the PQ will link with the requirements specification deliverable; the OQ will link with the company system specification deliverable; and the IQ will link with the technical design specification deliverable.

In addition, all the document inspections and technical review results for all the technologies used for the systems must be evaluated. A summary of nonconformances and deviations, their resolution and impact, must also be included.

For very large validation projects, the project report should reference (by title and document reference number) other documents satisfying protocol requirements. In smaller validation projects, the actual evidence can be incorporated into the project as appendices.

Documents associated with the system qualification and their results must be assembled and reviewed by appropriately qualified personnel following the review, by QA, and by the personnel appropriate for the level of criticality of the system.

The approval of the qualification reports provides confirmation that the computer system as a whole is fit for its purpose, and that all essential documentation is available. For computer systems controlling manufacturing equipment (e.g., process control systems), the approval of the qualification reports indicate that the computerized systems is released for Process/Product

Performance Qualification. On other computer systems, the approval of the reports releases the system to the user.

The test results for compliance with Part 11 must be included in, or referenced by the associated qualification report.

The project report summarizes the outcome of each activity performed to develop or maintain the computer system and the verification that critical checkpoints were reached during the entire development process. It also verifies that the quality development procedures identified in the project plan were complied with.

All verification and testing results obtained during the project must also be addressed in the project report. The approval of the project report enables the computer system to be released for use.

Conduct a Process/Product Performance Qualification[2]

The product/process PQ is documented verification that the process and/or the total process-related system perform as intended throughout their normal operating ranges.

In practice, this process ensures that the system (computer hardware, computer software, controlled equipment, interfaces, operating environment, operators, equipment, procedures, and so on), in its normal operating environment, produces acceptable quality product and that sufficient documentary evidence exists to demonstrate this assessment.

Traditionally, the qualification documents described above contain both specification information and instructions on testing.

Formal Release for Use

An appropriate signature sign-off, as defined in the validation plan, must be obtained prior to officially utilizing the system in its operating environment. Once the system is considered 'validated,' all applicable policies and procedures governing the operation of the validated computer system become effective. A formal written notification should be issued to the system users and to QA to officially inform them that the system is validated and can be used in operational environment.

After the computer system is put into operation and all appropriate validation documentation has been collected, as defined in the validation plan, the entire validation documentation package should be archived to ensure its safety and security.

Validation Process Deviations

A deviation is a completed event not conforming with the validation plan or validation procedures. It may include documentation errors, test results not meeting the acceptance criteria, or deviations to a procedure.

When a deviation takes place, the project manager and/or tester executing the protocol must

---

[2] *Note*: This activity is the culmination of computer systems controlling manufacturing equipment and is out of the scope of computer systems validation. This qualification may be performed by QA or R&D and only on manufacturing systems.

complete a test incident (or similar) report. This report describes the nature of the event, the corrective action(s) necessary to resolve the event, an assessment of the impact of the event on the test case or any related documentation, software and any subsequent test cases and/ or protocols.

The validation process deviations must be written in a procedure.

OPERATIONAL LIFE

Refer to Chapter 18.

RETIREMENT

The retirement of computer systems, which have been used in the development, clinical, manufacture, packaging, or marketing of regulated operations, is a critical process for a company. The purpose of the SLC 'retirement period' is to replace or eliminate the current computer technologies supporting a regulated operation, and to ensure the availability of the data generated by it for conversion, migration, or retirement.

It is a requirement of Part 11 that all data produced from electronic files is in human readable form and that true copies are available. In the case of data retirement, the planning must include the technologies needed to support the retired records, for the duration of the retention period required in the applicable regulations. If electronic records which have been retired need to be accessed in the future for review, the data and the application source code need to be readily accessible. The qualification of any conversion processes for the data prior to archive/retirement must pay special attention to these points.

A 'decommissioning' plan should be written prior to the retirement of a system to ensure continuity of service.

In some cases, a replacement system will be installed which can access the data records from the previous system, including historical data, to enable such data to be available in the format used by the new system. If necessary, a data conversion process must be carried out. Where system obsolescence forces the need to convert electronic records from one system or format to another, the conversion process must be qualified and its integrity verified. The electronic copies of electronic record files can be considered to be electronic record files provided that the data transfer processes have been verified. Services such as Digital Notary expedite the verification of the conversion process.

Following the successful conversion of the electronic records, and QA acceptance of the validity of the conversion, the new system may be authorized for use. The results of the validation must be fully documented and added to the validation file of the system that is being retired.

In other cases, the electronic records will not be converted for use by the new system. In this case, the obsolete system could be maintained as a legacy system, and utilized to access the electronic records. This approach can be expensive and is one that might still require the transfer of the legacy data to the new system or format at a later date, when maintenance becomes impractical.

During product research, production, and control of FDA regulated products, documentary evidence must be retained for a certain period, which prove the safety and efficacy of the product. When a computer system is to be retired from active use, the data from that system must be archived. Some regulations which cover FDA regulated products, and which require records to be retained, are listed in Table E–1.

Table E–1. Regulations Which Require the Retention of Records.

AREA	SECTION	REGULATION
GLP	Organization and personnel records	FDA 58.3
GLP	Specimen and data storage facilities	FDA 58.51
GLP	Storage and retrieval of records and data	FDA 58.190 & 58.195
GLP	Organization and personnel	OECD 1.1 & 1.2
GLP	Sampling and storage	OECD 6.1
GLP	Archive facilities	OECD 3.4
GLP	Characterization	OECD 6.2
GLP	SOPs	OECD 7.2
GLP	Storage and retrieval	OECD 10.1 & 10.2
GLP	Storage and retrieval with archive facilities	OECD 10.2
GCP	Supply and handling of investigational product(s)	ICH 5.14
GCP	Trial management, data handling, record keeping	ICH 5.5
GMP	Records and documentation	FDA 211.180
GMP	Personnel records	EU PIC 2
GMP	Documentation	EU PIC 4

The company may wish to keep other data produced during the course of an operation, e.g., data on products never sent for production. This would be a business decision for the company, and is not a regulatory requirement.

In addition to the retention of operating data in electronic format, the following list shows the supporting documentation that should be archived along with the data. The integrity of archived data is assured when the systems on which the data was produced have been validated, and the documentation relating to these systems has been assembled, as part of the validation file.

This is the key documentation to be retained for a validated system due for retirement:

- IQ, OQ, PQ protocols and reports (when applicable)
- Project and Quality Plan
- Process/Operation and requirements specification
- criticality analysis
- computer system specification
- system drawings
- system (hardware and software) configuration and design
- source code (if available)
- application software documentation.
- system testing
- development history and/or system development files
- system manuals
- user manuals
- technical manuals
- list of users trained to use the system and proof of the training records
- technical training manuals
- equipment inventory list
- inventory list (hardware and software)
- calibration list
- log book(s)

- history of changes to the system
- history of audits

When all computer systems decommissioning activities have been completed, the personnel responsible for performing the decommissioning activities may issue a memo that indicates the following:

- the reason for decommissioning the system
- the date that the system will be decommissioned
- the system documentation and electronic records retention period expiration date indicating a final date for which the system documentation must be retained, following system decommission
- signatures from the appropriate personnel
- signatures from the designated personnel involved in performing and verifying completion of the decommissioned activities

COMPUTER SYSTEM INCIDENTS

The malfunction or failure of system components, incorrect documentation, or improper operation that makes the proper use of the system impossible for an undetermined period, characterizes some of the incidents that can affect the correct operation of a computer system. These system incidents are also nonconformances.

In order to remedy problems quickly or to avoid system errors, appropriate procedural controls must be applied to (a) enable early identification of an imminent breakdown, and (b) initiate appropriate preventive action. The next section suggests a course of action to follow to identify, analyze, and record computer system incidents, and to implement and test the corrective action taken.

- A qualified person should conduct a root cause analysis in order to determine the nature of the malfunction or error. If it is determined that the malfunction is equipment related, the following process may be implemented.
  - The personnel responsible for managing the component causing the incident must:
    - ➢ complete a problem description form and include an initial error analysis
    - ➢ record the incident using an incident log
  - Check any stored data and, if applicable, reconstruct the data.

- If the error analysis demonstrates that a malfunction or failure of the software, or of a hardware component has occurred, then:
  - using a root cause analysis, identify and specify the modification needed to fix the malfunction
  - fill out the appropriate change control forms and submit these for approval. Include the root cause analysis report
  - if the system modification or replacement part does not constitute an emergency change, wait until the change control forms are approved
  - if the system modification or replacement part constitute an emergency change, a qualified person may make the necessary modification(s) immediately. In this case, after the modification have been made, the system can return to production and the following steps implemented, as appropriate

- perform a criticality and complexity assessment in order to determine the documentation requirements for the modification and to enable the approved system development lifecycle model to be followed
- if the system modification or replacement part is not an emergency, carry out the corrective actions identified
- if a test environment is available, test the software/hardware that have been modified before the system is returned to the operational environment
- if applicable, update the system documentation and/or user's manual
- if the user's manual was updated, conduct user training on the changes
- if the analysis demonstrates incorrect documentation
  - ➢ update the system documentation and/or user's manual
  - ➢ conduct user training on the revised documentation
- if the analysis demonstrates improper operation
  - ➢ conduct user refresher training on the correct operation of the system

• Release the system for use

If the analysis uncovers incorrect documentation or improper operation, the originator of the report, the area manager and the technical manager(s) must approve the incident report, including the recommendation made, and send it to the facility's quality assurance department for review and approval. The signed report should be returned to the originator, system administrator, or system owner for implementation. On completion of the implementation, the original report and associated documentation must be filed in the site document retention center, and/or with the change control documentation for the computer system.

If an observed incident constitutes an emergency, a qualified person may make changes immediately. In this case, after the changes have been implemented, the system can return to production. The area manager and the site quality assurance department must be notified of the changes(s) in writing before the close of business on the day after the incident, a formal incident report must be filed before the close of business on the day after the incident, and all other required processes and documentation must be completed. The processes may include (as appropriate): the review and update of the Requirements Traceability Matrix, an update of the requirements, computer system and technical design specification, deliverables, the development of integration tests, the modification and application of Installation Qualification and Operational Qualification in both the test environment and the production environment, the modification and application of Performance Qualification, the modification of manuals and training materials, the presentation of incremental user training, and the preparation of the Change Implementation Report.

If incidents are managed electronically, all necessary controls must be in place for the creation, maintenance, archiving, review, and approval of the incidents.

In both the paper-based and electronic-based environment the System Incident Report or its equivalent must be retained according to the company's record-retention policies after the incidents are closed (See Figure E–3).

SYSTEM LIFE CYCLE

Start → ... → End

Period	Concept	Development					Early Life, Maturity and Aging	Retirement
		Requirements	Design	Build	Test	Acceptance and Implementation		
Event	Recommendation	Project Initiation				Release for use		
Outputs	Defined need to automate or reengineer	Reqs. Spec. Validation Plan Risk Analysis; RFP; RFP Response Vendor Audit (if applicable); System Spec. Criticality and Complexity	Technical Design	Code Code reviews, inspections, audits	Unit and Integration Test Plans & Results; User Documentation System Documentation Training Documentation	Installation and checkout (SAT) Plans; IQ/OQ/PQ Plans and Sum. Docs.; Procedural Controls; Validation / Project Summary Report; Release Memo	Procedural controls and records. Performance, change control, periodic review records.	Decision to re-engineer or replace. Retirement Plan
			System Configuration (if applicable) →		In-process audits →	FAT →	Support Log	
Part 11 Considerations	Include Part 11	Include Part 11 in: • Reqs Spec, • Val Plan, • RFP, • Vendor audit • System Spec.	Build technical controls into system. Test technical controls in Unit, Integration and FAT.			Include procedural controls in SOPs. Test technical controls in SAT and OQ/PQ. Include Part 11 in VSR	Part 11 corrective, adaptive, and perfective maintenance. Assess new technologies as applicable.	Include Part 11 requirements in Retirement Plan

Figure E-3. Sample Development Activities and Work Products.

Appendix F

Administrative Procedures Mapped to Part 11[1]

ADMINISTRATIVE PROCEDURES NECESSARY TO GUARD DATA INTEGRITY, CONFIDENTIALITY, AND AVAILABILITY

Requirement	Implementation
Certification Evaluation to certify that the computer and network are appropriately secure.	
Chain of trust partner agreement If data is processed through a third party, the parties will be required to enter into a chain of trust, or similar, partner agreement.	
Contingency plan (all listed implementation features must be present) The contingency plan enables a system emergency to be responded to and includes: • Applications and data criticality analysis • Data backup plan • Disaster recovery plan • Plan for the emergency operating mode • Testing and revision	• Applications and data criticality • Data backup plan • Plan for the emergency operating mode • Testing and revision procedures
Formal mechanism for processing records Documented SOPs for the routine and non-routine receipt, manipulation, storage, dissemination, transmission, and/or disposal of information.	
Information accesses control (all listed implementation features must be present) Granting users different levels of access to information.	• Access authorization • Access establishment • Access modification
Internal audit An in-house review of the records of system activities (e.g., logins, file accesses, security incidents)	

[1] Department of Health and Human Services, Office of the Secretary, *45 CFR Part 142 Security and Electronic Signature Standards; Proposed Rule*, Federal Register 63 (155), 43242–43280, August 12, 1998.

Requirement	Implementation
Personnel security (all listed implementation features must be present) All personnel with access to health information must be authorized to do so after receiving the appropriate clearance.	• Supervision of maintenance personnel by an authorized, knowledgeable person • Maintenance of record of access authorizations • Operating, and in some cases, maintenance personnel have proper access authorization • Personnel clearance procedure • Personnel security policy/procedure • System users, including maintenance personnel, are provided with security training
Security configuration management (all listed implementation features must be present) The organization to implement the necessary measures, practices, and procedures for the security of information systems	• Documentation • Hardware/software installation and maintenance review and the testing of the security features • System inventory • Security testing • Virus checking
Security incident procedures (all listed implementation features must be present) Implement accurate and current security incident procedures. The reporting of security breaches, so that security violations are reported and handled promptly.	• Report procedures • Response procedures
Security management process (all listed implementation features must be present) Creating, administering, and overseeing procedures to ensure the prevention, detection, containment, and correction of security breaches.	• Risk analysis • Risk management • Sanction policy • Security policy
Termination procedures (all listed implementation features must be present) Procedures for the ending of an employee's employment or an internal/external user's access.	• Combination locks changed • Removal from access lists • Removal of user account(s) • Return of keys, tokens, or cards that allow access
Training (all listed implementation features must be present) Security training for all staff on the procedures that must be followed to ensure the protection of information.	• Awareness training for all personnel (including management) • Periodic security reminders • User education concerning virus protection • User education on the importance of monitoring logins, success/failure of system logins, and how to report discrepancies • User education in password management

PHYSICAL SAFEGUARDS NECESSARY TO GUARD DATA INTEGRITY, CONFIDENTIALITY, AND AVAILABILITY

Requirement	Implementation
Assigned security responsibility The assignment of responsibility for security to a specific individual or organization.	
Media controls (all listed implementation features must be present) This area is related to the procedures that govern the receipt and removal of hardware/software/data.	• Access control • Accountability (tracking mechanism) • Data backup • Data storage • Disposal
Physical access controls (limited access) (all listed implementation features must be present) This area is related to the procedures for limiting physical access to an entity while ensuring that properly authorized access is allowed.	• Disaster recovery • Emergency operating mode • Equipment control (into and out of site) • Facility security plan • Procedures for verifying access authorizations prior to physical access • Maintenance records • Need-to-know procedures for personnel access • Sign-in for visitors and their escorts, if appropriate • Testing and revision
Policy/guideline on workstation use Procedures to maximize the security of information (e.g., logging off before leaving a terminal unattended).	
Secure workstation location This area is related to the physical safeguards that eliminate or minimize the possibility of unauthorized access to information.	
Security awareness training This is a requirement for all employees and contractors to understand their security responsibilities, based on job responsibilities in the organization, and to make security a part of their daily activities.	

TECHNICAL SECURITY SERVICES NECESSARY TO GUARD DATA INTEGRITY, CONFIDENTIALITY, AND AVAILABILITY

Requirement	Implementation
Access control *Note*: **Emergency access procedures must be present, and at at least one of the following three implementation features must be present: context-based access, role-based access, user-based access. The use of encryption is optional.** The objective is to implement an access control service that would restrict access to resources and that would only allow access by privileged entities.	• Context-based access • Emergency access procedures • Encryption • Role-based access • User-based access
Audit controls Put in place audit control mechanism to record and examine system activity.	
Authorization control (at least one of the listed implementation features must be present) See access control above.	• Role-based access • User-based access
Data authentication To provide evidence that the data possessed by the system has not been altered or destroyed in an unauthorized manner. Examples of how data authentication may be assured include the use of an authentication code, or digital signature.	
Entity Authentication • The following implementation features must be present: – Automatic logoff – Unique user identification • In addition, at least one of the other listed implementation features must be present to verify that an entity is who it claims to be.	• Automatic logoff • Biometrics • Password • PIN • Telephone callback • Token • Unique user identification

TECHNICAL SECURITY MECHANISMS NECESSARY TO GUARD AGAINST UNAUTHORIZED ACCESS TO DATA THAT IS TRANSMITTED OVER A COMMUNICATIONS NETWORK

Requirement	Implementation

Communications/network controls
• The following implementation features must be present:
   – integrity controls
   – message authentication
• If communications or networking is employed, one of the following implementation features must be present:
   – access controls
   – encryption
• In addition, if using a network, the following implementation features must be present:
• alarm
• audit trail
• entity authentication
• event reporting
	• access controls
	• alarm
	• audit trail
	• encryption
	• entity authentication
	• event reporting
	• integrity controls
	• message authentication

Sample Audit Checklist for a Closed System

ELECTRONIC SIGNATURE AND ELECTRONIC RECORDS

AUDIT CHECKLIST FOR A CLOSED SYSTEM

General Vendor Checklist

<table>
<tr><td colspan="2">Appendix G objective:</td></tr>
<tr><td colspan="2">The purpose of the following checklist is to help to determine if a computer system complies with the FDA Rule 21 CFR 21 Part 11 for electronic records and electronic signatures. This audit questionnaire applies to systems that meet the definition of a 'closed' system as defined in Section 11.3 (b)(4) of the rule and which do not utilize biometrics identification methods.</td></tr>
</table>

Section	Question	Answer Yes No N/A	Comments
11.10(b)	Are data entries stored in the database tables in an encrypted form?	☐ ☐ ☐	
11.10(b)	Can data entries be printed directly from the database, then read by a human without a key?	☐ ☐ ☐	
11.10(b)	Can records be output in common formats such as ASCII, VSAM, XLS, PDF, etc.?	☐ ☐ ☐	
11.10(e) 11.50	Is every data or signature entry recorded in the database with a date and time stamp?	☐ ☐ ☐	
11.10(e)	Is an entry that has been made and then later corrected retained in the system?	☐ ☐ ☐	
11.10(e)	Is a prior entry that has been corrected identified as superseded?	☐ ☐ ☐	

Section	Question	Answer			Comments
		Yes	No	N/A	
11.10(e)	If an entry is made by the operator, but is not saved due to a warning about its potential incorrectness retained in the system?	☐	☐	☐	
11.10(e)	Does the system include in its audit trail operator actions that are not based upon a data entry or signature, (e.g., when an operator triggers an action, responds to a message, etc.?	☐	☐	☐	
11.10(d)	Does the system limit who may access the system functions?	☐	☐	☐	
11.10(g)	Does the system restrict individuals to specific documents?	☐	☐	☐	
11.10(g) 11.10(h)	Does the system restrict individuals to specific workstations, functions, or documents?	☐	☐	☐	
11.10(f)	Does the system have the capability to control the sequence in which users may view screens?	☐	☐	☐	
11.10(f)	Does the system have the capability to detect when an entry occurs outside of the normal sequence?	☐	☐	☐	
11.10(h)	Does the system record the location (node) of the workstation where each entry was made?	☐	☐	☐	
11.50(b)	Does the database table for each entry show the unencrypted name of the entry maker?	☐	☐	☐	
11.50(a)	Does the system store the meaning or description of each signature	☐	☐	☐	
11.50(b)	Are the details of the signing (e.g., signer, date, time, meaning) included on every display and printout?	☐	☐	☐	
11.70	Does the system have an absolute method of linking each signature to its corresponding entry?	☐	☐	☐	
11.200(a)	Does each signature entry have two separate methods for identifying who is making that signature, and are both of these supplied by the operator every time he signs his name?	☐	☐	☐	

Section	Question	Answer Yes	No	N/A	Comments
11.10(c)	Can inactive records that are still within the legal retention period be readily retrieved, completely intact, with their full audit trail?	☐	☐	☐	
11.10(c)	Does the system have a security scheme to protect database(s) from accidental or deliberate damage?	☐	☐	☐	
11.10(k)	Is there a written procedure in place for the control, change, distribution, and destruction of system documentation?	☐	☐	☐	
11.10(k)	Are written records kept of the copies and possession of copies for all system documentation?	☐	☐	☐	
11.10(k)	Does every system documentation change have approvals and a document change history?	☐	☐	☐	
11.100 11.300(a)	Is there any possibility that two people could share the same signature components at any time in the past, present, or future?	☐	☐	☐	
11.100	Is there any possibility that two or more people could share the same name (e.g., for a database table)?	☐	☐	☐	
1.100(b)	Is there a written form for the assignment of electronic signatures?	☐	☐	☐	
1.100(b)	Does the signature assignment procedure require positive identification of the person to whom the signature codes are given?	☐	☐	☐	
11.200	Is there any way that one person could learn and/or use the signature codes of another person?	☐	☐	☐	
11.300(d)	Does the system have a method to detect the unauthorized use of signature codes?	☐	☐	☐	
11.10(a)	Is there a current system validation for all system software and hardware components?	☐	☐	☐	

Section	Question	Answer Yes	No	N/A	Comments
11.10(a)	Does the system have a method to detect the alteration of any record, even if the alteration was done directly to the database?	☐	☐	☐	
11.300(b)	Are signature codes periodically changed or verified?	☐	☐	☐	
11.300(b)	Are there records in existence that confirm that signature codes are periodically changed or verified?	☐	☐	☐	
11.300(c)	Is there a written procedure in place for the replacement of any objects (tokens) that are used for signature purposes when they become damaged or lost?	☐	☐	☐	
11.300(c)	Are there records in existence that confirm that the procedure for the replacement of objects (tokens) is being followed?	☐	☐	☐	
11.300(e)	Is there a procedure in place for the periodic testing of any objects (tokens) that are used for signature purposes?	☐	☐	☐	
11.300(e)	Are there records in existence that confirm that the procedure for the periodic testing of objects (token) is being followed?	☐	☐	☐	
11.10(i)	Are there records or statements of qualifications for all of the persons who developed, installed, and/or validated the system?	☐	☐	☐	
11.10(i)	Is there a user-training program with course materials for all categories of system user?	☐	☐	☐	
11.10(i)	Are there training records/assessments of the user training received for every system user?	☐	☐	☐	
11.100(c)	Has the company/location notified the FDA that it is using electronic signatures?	☐	☐	☐	
11.100(c)	Has the company determined the method that it will use to supply additional testimony as to the validity of a specific signature?	☐	☐	☐	

Appendix H

Computer Systems Regulatory Requirements

This appendix discusses the main FDA regulatory requirements and guidelines applicable to computer systems performing regulated operations and their interpretation. The following regulations are covered:

- Guidelines on General Principles of Process Validation, May 1987
- Human and veterinary products GMP Regulations (21 CFR 210 and 211)
- Nonclinical laboratory GLP Regulations (21 CFR 58)
- Medical Devices GMP Regulations (21 CFR 820)
- Food (21 CFR 110 and others)

GUIDELINES ON GENERAL PRINCIPLES OF PROCESS VALIDATION

This guideline, developed in 1987, outlines the general principles considered by the FDA as 'acceptable elements of process validation for the preparation of human and animal drug products, and for medical devices.' Our analysis of the elements impacting computer systems will concentrate on Section VIII Elements of Process Validation. The connection between the Process Validation guideline and computer systems performing regulated operations is established in CPG 7132a.11

VIII (A) – PROSPECTIVE VALIDATION

Prospective validation includes considerations that need to be made before a firm introduces an entirely new product, or when there is a change in the manufacturing process that may affect the product's characteristics, such as uniformity and identity.

ANALYSIS

1 All computer systems performing regulated operations should be validated; the depth and scope of validation depends on the type of software and the complexity and criticality of the application.
2 Validation should be performed to verify the identifiable performance requirements of an application.
3 Before a system is brought into use, it should be thoroughly tested and confirmed as being capable of achieving the desired results and performance.
4 All new computer systems must be prospectively validated before they are brought into operational use.

VIII (A)(1) – EQUIPMENT AND PROCESS

The equipment and process(es) should be designed and/or selected so that product specifications are consistently achieved. This should be done with the participation of all appropriate groups concerned with assuring product quality, e.g., engineering design, production operations, and quality assurance personnel.

ANALYSIS

1   Attention should be paid to the siting of computer hardware in suitable conditions where extraneous factors cannot interfere with the system.
2   Based on the identification of operational functions, design and proper computer performance, relevant computer hardware technologies can be selected. Depending on the available technology and its cost, automated functions can be assigned to the computer hardware and/ or software.
3   A set of design documentation, including as-built drawings, should be maintained for computers, infrastructure, and instrumentation.
4   Integration of hardware and the development and testing of hardware and software should be performed under a quality assurance system. For computer technology suppliers and developers, the quality assurance system must be agreed upon and defined in a contract between the supplier and customer.

VIII (A)(1)(A) – EQUIPMENT: INSTALLATION QUALIFICATION

Installation qualification studies establish a degree of confidence that the process equipment and ancillary systems are capable of consistently operating within established limits and tolerances. After process equipment is designed or selected, it should be evaluated and tested to verify that it is capable of operating satisfactorily within the operating limits required by the process. This phase of validation includes examination of equipment design; determination of calibration, maintenance, and adjustment requirements; and the identification of critical equipment features that could affect the process and product. Information obtained from these studies should be used to establish written procedures covering equipment calibration, maintenance, monitoring, and control.

In assessing the suitability of a given piece of equipment, it is usually insufficient to rely solely upon the representations of the equipment supplier, or upon experience in producing some other product. Sound theoretical and practical engineering principles and considerations are the first steps in the assessment.

It is important that equipment qualifications simulate actual production conditions, including those that are 'worst-case' situations. Tests and challenges should be repeated a sufficient number of times to assure reliable and meaningful results. All acceptance criteria must be met during the test or challenge. If any test or challenge shows that the equipment does not perform within its specifications, an evaluation should be performed to identify the cause of the failure. Corrections should be made and additional test runs performed, as needed, to verify that the equipment performs within specifications.

The observed variability of the equipment between and within runs can be used as a basis for determining the total number of trials selected for the subsequent performance qualification studies of the process. Once the equipment configuration and performance characteristics are established and qualified, they should be documented. The installation qualification should include a review of pertinent maintenance procedures, repair parts lists, and calibration methods for each piece of equipment. The objective is to assure that all repairs can be performed in a manner that will not affect the characteristics of material processed after the repair.

In addition, special post-repair cleaning and calibration requirements should be developed to prevent inadvertent manufacture of a nonconforming product. Planning during the qualification phase can prevent confusion during emergency repairs that could lead to the use of the wrong replacement part.

ANALYSIS

1   Assurance of computer systems reliability is achieved by the execution of a quality plan and by testing during the software development process. This involves unit code testing and integration testing in accord with the principles specified in ISO 9000–3 and IEEE 1298 (PIC/S).

2   Appropriate IQ, OQ and PQ tests should demonstrate the suitability of computer hardware and software to perform their assigned tasks and to ensure their proper performance in an operational environment. While IQ, OQ, and PQ terminology has served its purpose well, and is one of many legitimate ways to organize dynamic testing in the operational environment for the FDA-regulated industries, this terminology may not be well understood among many software professionals. However, organizations performing regulated operations must be aware of these differences in terminology as they request and provide information regarding computer systems.

3   The software is a critical component of a computer system. The user should take all reasonable steps to ensure that it has been produced under a quality assurance system.

4   Written procedures should be available to:
   • operate the system under normal and abnormal situations
   • detect and record errors, and enable corrections to be made
   • restore a system to an operational condition including data recovery
   • authorize, carry out, and record changes
   • enable, when appropriate, electronic signatures

Refer to Chapter 12 and Appendix F.

VIII (A)(2) – SYSTEM TO ASSURE TIMELY REVALIDATION

There should be a quality assurance system in place that requires revalidation whenever there are changes in packaging, formulation, equipment, or processes that could impact on product effectiveness or product characteristics, and whenever there are changes in product characteristics. Furthermore, when a change of raw materials supplier occurs, the manufacturer should consider subtle, potentially adverse differences in the raw material characteristics. A determination of adverse differences in raw material indicates a need to revalidate the process.

One way of detecting the kind of changes that should initiate revalidation is the use of tests and methods of analysis capable of measuring characteristics that may vary. Such tests and methods usually yield specific results going beyond the mere pass/fail basis, thereby detecting variations within product and process specifications and allowing determination of whether a process is slipping out of control.

The quality assurance procedures should establish the circumstances under which revalidation is required. These may be based upon equipment, process, and product performance observed during the initial validation challenge studies. It is desirable to designate individuals who have the responsibility to review product, process, equipment, and personnel changes to determine if and when revalidation is warranted.

The extent of revalidation will depend upon the nature of the changes and how they impact upon different aspects of previously validated production. It may not be necessary to revalidate a process from scratch merely because a given circumstance has changed. However, it is important to carefully assess the nature of the change to determine potential ripple effects, and what needs to be considered as part of revalidation.

ANALYSIS

1   All changes made to the computer system should be formally authorized, documented, and tested. Records should be kept of all changes including modifications and enhancements made to the hardware, software and any other critical component of the system to demonstrate that the final system has been maintained in a validated state.
2   When appropriate, regression testing (refer to Chapter 9) must be performed. The depth and scope of revalidation depends on the type of software and the complexity, and criticality of the application.
3   All revalidations must be prospective, before they are brought into operational use.

VIII (A)(3) – DOCUMENTATION

It is essential that the validation program is documented and that the documentation is properly maintained. Approval and release of the process for use in routine manufacturing should be based upon a review of all the validation documentation, including data from the equipment qualification, process performance qualification and product/package testing to ensure compatibility with the process.

For routine production, it is important to adequately record process details (e.g., time, temperature, equipment used) and to record all changes that have occurred. A maintenance log can be useful in performing failure investigations concerning a specific manufacturing lot. Validation data (along with specific test data) may also determine expected variance in product or equipment characteristics.

ANALYSIS

1   Refer to Chapter 2, Key Validation Elements.

GOOD MANUFACTURING PRACTICE REGULATIONS RELATED TO COMPUTER SYSTEMS VALIDATION

The GMP regulations have been in existence since 1963. The current regulations are from 1978 (43 FR 45077), and amended in 1995 (60 FR 4091). The proposed GMP regulation from 1996, includes Section 211.220(a) that would require the validation of computer systems performing manufacturing-regulated operations. In addition to section 211.220(a), several other sections apply to computer systems and to responsible individuals.

SUB-PART B – ORGANIZATION AND PERSONNEL

211.25 Personnel Qualifications

(a) Each person engaged in the manufacture, processing, packing, or holding of a drug product shall have education, training, and experience or any combination thereof, to perform the assigned functions. Training shall be in the particular operations that the employees perform and in current good manufacturing practice.

(b) Each person responsible for supervising the manufacture, processing, packing, or holding of a drug product shall have the education, training, and experience or any combination thereof, to perform assigned functions in such a manner as to provide assurance that the drug product has the safety, identity, strength, quality, and purity that it purports or is represented to possess.

ANALYSIS

1 Appropriate training must be provided for personnel involved with computer systems and must also include training in GMPS.

2 The documentation of qualifications including the education, training, experience, or any combination thereof must be available for developers, users, and support personnel.

SUB-PART C – BUILDING AND FACILITY

211.42 Design and Construction Features

(a) Any building or buildings used in the manufacture, processing, packing, or holding of a drug product shall be of suitable size, construction and location, to facilitate cleaning, maintenance, and proper operations.

ANALYSIS

The area allocated to the computer system must be suitable for the proper operation of the system.

SUB-PART D – EQUIPMENT

211.63 Equipment Construction

Equipment used in the manufacture, processing, packing, or holding of a drug product shall be of appropriate design, adequate size, and suitably located to facilitate operations for its cleaning and maintenance.

ANALYSIS

1   Computer hardware must be properly specified to meet the requirements for its intended use, and the amount of data it must handle.
2   Environmental controls, electrical requirements, electromagnetic 'noise' control, and others should be considered when determining a location for the hardware.
3   The location of the hardware must allow access for maintenance, as required.

211.68 Automatic, Mechanical, and Electronic Equipment

(a) Automatic, mechanical, or electronic equipment or other types of equipment, including computers, or related systems that will perform a function satisfactorily, may be used in the manufacture, processing, packing, and holding of a drug product. If such equipment is so used, it shall be routinely calibrated, inspected, or checked according to a written program designed to assure proper performance. Written records of those calibration checks and inspections shall be maintained.

(b) Appropriate controls shall be exercised over computer or related systems to ensure that changes in master production and control records or other records are instituted only by authorized personnel. Input to and output from the computer or related system of formulae or other records or data shall be checked for accuracy. The degree and frequency of input/output verification shall be based on the complexity and reliability of the computer or related system. A backup file of data entered into the computer or related system shall be maintained except where certain data, such as calculations performed in connection with laboratory analysis, are eliminated by computerization or other automated processes. In such instances a written record of the program shall be maintained along with data establishing proper performance. Hard copy or alternative systems, such as duplicates, tapes, or microfilm, designed to assure that backup data are exact and complete and that it is secure from alteration, inadvertent erasures, or loss shall be maintained.

ANALYSIS

1   There must be a written program detailing the maintenance of the computer system (i.e., hardware maintenance manual). The maintenance of the computer, including periodic scheduled maintenance and breakdown maintenance, must be documented.
2   There must be a system to control changes to both the hardware and the software. Changes must only be made by authorized individuals following an appropriate review and approval of the change.
3   There must be documented checks of the system inputs and outputs for accuracy (refer to CPG 7132a.07, 'I/O Checking'). According to this CPG, computers I/Os are to be tested for data accuracy as part of the computer system qualification and, after the qualification, as part of the computer system's ongoing verification program. This CPG and the proposed 1996 GMP changes in Part 211.68(b) are considered to be an implicit requirement to validate all computer systems covered under the GMPS. Proposed 211.220(a) and Part 11.10(a) explicitly require validation of computer systems.
4   Computer files/data must be backed up either by hard copy, microfilm, or magnetic media. Computer files/data must be maintained electronically.
5   There must be validation data and a written record of the programs used when backup data is eliminated by computerization.

SUB-PART F – PRODUCTION AND PROCESS CONTROLS

211.100 – Written Procedures: Deviations

(a) There shall be written procedures for production and process control designed to assure that the drug products have the identity, strength, quality, and purity they purport or are represented to possess. These written procedures, including any charges, shall be ... reviewed and approved by the quality control unit.

ANALYSIS

1   There must be written procedures covering the operation of computer systems.
2   The Quality Assurance unit must review and approve procedural controls for computer systems.

SUB-PART J – RECORDS AND REPORTS

211.180 General Requirements

(a) Any production, control, or distribution record ... specifically associated with a batch of drug product shall be retained for at least one year after the expiration date of the batch ... .

ANALYSIS

1   Electronic records must be maintained electronically. In the case of electronic batch records, they must be retained, as a minimum, for at least one year after the expiration date of the batch.
2   The retention periods for supporting documentation such as maintenance records, change control records and project documentation are as follows:
   (a) obsolete system documentation — one year after the expiration date of the last batch supported by the legacy function or system;
   (b) retired system documentation — one year after the expiration date of the last batch supported by the retired functionality/ system; current documentation — retain permanently until (a) or (b) above apply.

GOOD LABORATORY PRACTICE REGULATIONS RELATED TO COMPUTER SYSTEMS VALIDATION

The GLP regulations have been in existence since 1978. From 1984 the FDA began applying these regulations to computer systems, their development, use, and support. The FDA has stated that hardware is considered to be 'equipment' and software is considered to be 'records' under the regulations. This means that computer system development documentation and procedures are now subject to FDA inspection.

The following sections provide an interpretation of how the GLPs can be applied to computer systems and responsible individuals (note that the word 'systems' may be substituted for 'studies' in the regulations):

Part 58 – Good Laboratory Practice For Nonclinical Laboratory Studies

Computer systems are often used in nonclinical laboratory studies to record, analyze, store, and summarize data. In this context, the computer systems are subject Good Laboratory Practice (GLP) regulations.

Compliance with these regulations is intended to ensure the quality and integrity of the safety data that is filed.

SUB-PART B – ORGANIZATION AND PERSONNEL

58.29 Personnel

(a) Each individual engaged in the conduct of ... a nonclinical laboratory study shall have education, training, and experience, or a combination thereof, to enable that individual to perform the assigned function.
(b) Each testing facility shall maintain a current summary of training and experience and job description for each individual engaged in ... the conduct of a laboratory study.

ANALYSIS

1 Appropriate training must be provided for personnel involved in the computer system development, use, and support.
2 Maintain current job descriptions, training records, and a summary of the education received and qualification gained (e.g., a Curriculum Vitae).

58.31 Testing Facility Management

For each nonclinical laboratory study, testing facility management shall ... (c) assure that there is a quality assurance unit ... (e) ensure that personnel, resources, equipment, materials, and methodologies are available as scheduled, (f) assure that personnel clearly understand the functions they are to perform, (g) assure that any deviations from these regulations reported by the quality assurance unit are communicated ... and corrective actions are taken and documented.

ANALYSIS

1 Management is responsible for establishing and maintaining an effective GLP compliance program.
2 Management assures that:
   (a) personnel, facilities, equipment, materials, and methodologies are available as scheduled.
   (b) responsible personnel are available as scheduled and they clearly understand and carry out the functions that they are to perform.
   (c) Quality Assurance is established, and all deviations are reported, corrected, and documented.

58.35 Quality Assurance Unit

(a) Testing facility shall have a quality assurance unit ... who shall be responsible for monitoring each study to assure management that the facilities, equipment, personnel, methods, practices, records and controls are in conformance with the regulations ... the quality assurance unit shall be entirely separate from and independent of the personnel engaged in the direction and conduct of that study.

(b) The quality assurance unit shall:

Inspect each phase of a nonclinical laboratory study periodically and maintain written and properly signed records ... showing the date of the inspection, the study inspected, the phase ... of the study inspected, the person performing the inspection, findings and problems, action recommended and taken to resolve existing, problems, and any scheduled date for re-inspections ... any significant problems, which are likely to affect study integrity found during the course of an inspection, shall be brought to the attention of ... management immediately.

Periodically submit to management ... written status reports ... noting any problems and the corrective actions taken.

Determine that no deviations from approved protocols or standard operations procedures were made without proper authorization and documentation.

Prepare and sign a statement ... which shall specify the dates inspections were made and the findings reported to management ... .

(c) The responsibilities and procedures applicable to the quality assurance unit the records maintained ... and the method of indexing such records shall be in writing and shall be maintained. These items including inspection dates, the study inspected, and the name of the individual performing the inspection shall be made available for inspection to authorized employees of the Food and Drug Administration.

(d) A designed representative of the Food and Drug Administration shall have access to the written procedures established for the inspection and may request testing the facility management to certify that inspections are being implemented, performed, documented, and followed up ... .

(e) All records maintained by the quality assurance unit shall be kept in one location at the testing facility.

ANALYSIS

Quality Assurance must be established for the following:

1 Monitoring systems to assure management that the facilities, equipment, personnel, methods, practices, records and controls are in conformance with the regulations.

2 Exist entirely separate from and independent of the personnel engaged in developing and use of laboratory computer systems.

3 Inspect systems at intervals adequate to ensure the integrity of the system and its data.

4 Determine that procedural controls have been written, and that no deviations from them were made without proper authorization and documentation.

5 Retain documented and properly signed records of each inspection, indicating the date, what was inspected, responsible individual, findings/problems, resolution, and the scheduled date for reinspection.

6 Report to management immediately any significant problems likely to affect study integrity.

7 Submit periodic written reports of findings to management, and include any deviations from the regulations and the corrective action taken.

8   Maintain QA procedural controls for responsibilities and procedures, and make them available for inspection by the FDA.

SUB-PART C – FACILITIES

58.41 General

Each testing facility shall be of suitable size, construction, and location to facilitate the proper conduct of nonclinical laboratory studies.

ANALYSIS

Each area defined for housing the computer system must be suitable for the proper operation of the system.

SUB-PART D – EQUIPMENT

58.61 Equipment Design

... electronic equipment used in the generation, measurement, or assessment of data and equipment used for facility environmental control shall be of appropriate design and adequate capacity to function according to the protocol and shall be suitably located for operation, inspection, cleaning, and maintenance.

58.63 Maintenance and Calibration of Equipment

(a) Equipment shall be adequately inspected, cleaned and maintained. Equipment used for the generation, measurement, or assessment of data shall be adequately tested, calibrated, and/or standardized.
(b) The written standard operating procedures ... shall set forth in sufficient detail the methods, material, and schedules to be used in the routine inspection, cleaning, maintenance, testing, calibration and/or standardization of equipment, and shall specify remedial action to be taken in the event of failure or malfunction of equipment. The written standard operating procedures shall designate a person for the performance of each operation, and copies of the standard operating procedures shall be made available to laboratory personnel.
(c) Written records shall be maintained for all inspection, maintenance, testing, calibrating and/or standardizing operations. These records, containing the date of the operation, shall describe whether the maintenance operations were routine and followed. Written records shall be kept of nonroutine repairs performed on equipment as a result of failure and malfunction. Such records shall document the nature of the defect, how and when the defect was discovered, and any remedial action taken in response to the defect.

ANALYSIS

Computer systems are composed of electronic equipment used in the generation, measurement or assessment of data.

1 Computer systems must be appropriately designed and have adequate capacity to function according to their specifications, and must be suitably located for evaluation and maintenance.
2 Documented development methodologies and programming standards are required.
3 Computer systems must be adequately inspected, tested, calibrated, and maintained.
4 Documented records of routine maintenance, testing, and calibration are required.
5 Procedural controls shall define how, when, and who is responsible for the system's inspections, testing, and maintenance (routine and nonroutine).
6 Documentation should be retained for nonroutine repairs as a result of failure and malfunction and should include the problem, how and when it was found and the remedial action taken.

SUB-PART E – TESTING FACILITIES

58.81 Standard Operating Procedures

(a) 'A testing facility shall have standard operating procedures in writing setting forth non-clinical study methods that management is satisfied are adequate to insure the quality and integrity of the data generated in the course of a study.'
'All deviations ... from standard operating procedures shall be authorized ... and shall be documented.'
(d) 'A historical file of standard operating procedures, and all revisions thereof, including the dates of such revisions, shall be maintained.'

ANALYSIS

1 Provide documented procedural controls.
2 Deviations in established procedural controls need to be properly authorized and documented.
3 Historical files of procedural controls and all revisions are to be maintained.
Note: Procedural controls for computer applications may include: development, use, and support of systems, data handling, storage and retrieval, maintenance, repair, calibration of equipment, management and personnel responsibilities, record retention, training.

SUB-PART G – PROTOCOL FOR AND CONDUCT OF A NONCLINICAL LABORATORY STUDY

58.130 Conduct of a Nonclinical Laboratory Study

(e) 'All data generated during the conduct of a nonclinical laboratory study ... shall be dated on the day of entry ... the individuals responsible for direct data input shall be identified at the time of data input. Any change in computer entries shall be made so as not to obscure the original entry, shall indicate the reason for a change and shall be dated and the responsible individuals shall be identified.

ANALYSIS

Data must be identified (i.e., via an audit trail) with the following:

1   The individuals responsible for direct data entry.
2   The date and time of data input.
3   Documentation for all data corrections and including the original entry, reason for the change, date and responsible individual.

Note: Data may include computer printouts, magnetic media, and recorded data from automated instruments that are the result of the original observations, and activities of a non-clinical laboratory study necessary for the reconstruction and evaluation of the study report (electronic copies of electronic record files can be considered electronic record files provided that the data transfer processes have been verified).

SUB-PART J – RECORDS AND REPORTS

58.190 Storage and Retrieval of Records and Data

All raw data, documentation ... generated as a result of a nonclinical laboratory study shall be retained.

There shall be archives for orderly storage and expedient retrieval of all raw data, documentation ... Conditions of storage shall minimize deterioration of the documents ... in accordance with the requirements for the time period of their retention ... .

(c) Only authorized personnel shall enter the archives.

ANALYSIS

1   Raw data, system records, and documentation must be retained according to the periods stated in the regulations.
2   The data archive should be in a secure, limited access area and designed to preserve the data, records and documents.
3   Data, system records, and documentation should be archived in a way that permits expedient retrieval.

Note: Raw data, system records, and documentation may be retained as hard copy, on magnetic storage media, or microforms (microfiche).

58.195 Retention of Records

(b) '... documentation, records, raw data ... shall be retained for which ever of the following periods is the shortest':

(1) A period of at least 2 years following the date on which an application for research or marketing permit, in support of which the results of the nonclinical laboratory study were submitted, is approved by the Food and Drug Administration.

(2) A period of at least 5 years following the date on which the results of the nonclinical laboratory study are submitted to the FDA in support of an application for a research or marketing permit.

(c) The ... records of quality assurance inspections ... shall be maintained by the quality assurance unit as an easily accessible system of records for the period of time specified in ... this section.

(d) Summaries of training and experience and job descriptions ... may be retained ... along with all other ... employment records for the length of time specified in ... this section.

(f) Records and reports of the maintenance and calibration and inspection of equipment ... shall be retained for the length of the time specified in paragraph (b).

ANALYSIS

1   Support documentation such as maintenance records, change control records, and project documentation should be retained as required by the GLPs. Refer to 58.195. This requirement is applicable to paper-based and electronic-based records.

In addition, Quality Assurance records, summaries of training received, and job descriptions must be kept in permanent storage.

DEVICE GMP REGULATIONS (21 CFR 820) RELATED TO COMPUTER SYSTEMS VALIDATION

The device GMP regulations were published in 1978 (43FR 31508) and subsequently revised in 1979 (44FR 75628), 1988 (53FR 11253), 1990 (55FR 11169) and 1996 (61FR 52601). These revised regulations require the validation of computer systems. Systems covered by device GMP regulations include any system that directly or indirectly impacts the safety, effectiveness, or quality of materials, components, or the finished device.

SUB-PART A – GENERAL PROVISIONS

820.5 Quality System

(a) Each manufacturer shall establish and maintain a quality system that ensures that the requirements of this part are met, and that devices produced are safe, effective, and otherwise fit for their intended uses.

ANALYSIS

1   There must be a quality system in place appropriate to the specific device manufactured.

2   The quality system must include provisions for assuring the validation of all computer systems used in the manufacture of, or contained within, the finished device.

SUB-PART B – QUALITY SYSTEM REQUIREMENTS

820.20 Management Responsibility

(a) *Quality policy*. Management with executive responsibility shall establish its policy and objectives for, and commitment to, quality. Management with executive responsibility shall ensure that the quality policy is understood, implemented, and maintained at all levels of the organization.

(b) *Organization*. Each manufacturer shall establish and maintain an adequate organizational structure to ensure that devices are designed and produced in accordance with the requirements of this part.

   (1) *Responsibility and authority*. Each manufacturer shall establish the appropriate responsibility, authority, and interrelation of all personnel who manage, perform and assess work affecting quality, and provide the independence and authority necessary to perform these tasks.

   (2) *Resources*. Each manufacturer shall provide adequate resources, including the assignment of trained personnel, for management, performance of work, and assessment activities, including internal quality audits, to meet the requirements of this part.

   (3) *Management representative*. Management with executive responsibility shall appoint, and document such appointment of, a member of management who, irrespective of other responsibilities, shall have established authority over and responsibility for: (i) Ensuring that quality system requirements are effectively established and effectively maintained in accordance with this part; and (ii) Reporting on the performance of the quality system to management with executive responsibility for review.

(c) *Management review*. Management with executive responsibility shall review the suitability and effectiveness of the quality system at defined intervals and with sufficient frequency according to established procedures to ensure that the quality system satisfies the requirements of this part and the manufacturer's established quality policy and objectives. The dates and results of quality system reviews shall be documented.

(d) *Quality planning*. Each manufacturer shall establish a quality plan that defines the quality practices, resources and activities relevant to devices that are designed and manufactured. The manufacturer shall establish how the requirements for quality will be met.

(e) *Quality system procedures*. Each manufacturer shall establish quality system procedures and instructions. An outline of the structure of the documentation used in the quality system shall be established where appropriate.

ANALYSIS

1 The quality management system should be the responsibility of a designated individual(s) who does not have direct responsibility for the performance of a manufacturing operation.

2 The designated individual(s) responsible for the quality management system must have procedures that ensure the proper review of computer system validation documents. This function should also include the approval or rejection of all computer hardware/software relating to the device.

3 The individual(s) responsible for the quality management system must also ensure that adequate procedures are in place for identifying, recommending or providing solutions for quality assurance problems relating to computer systems, and for verifying of the implementation of such solutions.

4   There must be planned and periodic audits of the quality management system by trained personnel who do not have direct responsibility for the area being audited. The results of these audits must be documented in written reports that are submitted to the appropriate management.

820.25 Personnel

(a) *General*. Each manufacturer shall have sufficient personnel with the necessary education, background, training, and experience to assure that all activities required by this part are correctly performed.
(b) *Training*. Each manufacturer shall establish procedures for identifying training needs and to ensure that all personnel are trained to adequately perform their assigned responsibilities.

ANALYSIS

1   Appropriate training must be provided for personnel involved with computer systems.
2   Quality assurance personnel must have, or be given specific training in, the auditing of SLC processes and the identification of problems relating to the validation of computer systems.

SUB-PART G – PRODUCTION AND PROCESS CONTROLS

820.70 Production and Process Controls

(c) *Environmental controls*. Where environmental conditions could reasonably be expected to have an adverse effect on product quality, the manufacturer shall establish and maintain procedures to adequately control these environmental conditions. Environmental control system(s) shall be periodically inspected to verify that the system, including necessary equipment, is adequate and functioning properly. These activities shall be documented and reviewed.

ANALYSIS

1   If environmental controls are required to ensure the proper operation of a computer system, they must be periodically inspected and these inspections must be documented.
2   There must be documented periodic inspections (i.e., audits) to ensure adherence to the hardware maintenance schedule.

CLEANING AND SANITATION

820.70(d) Personnel

Each manufacturer shall establish and maintain requirements for the health, cleanliness, personal practices, and clothing of personnel if contact between such personnel and product or environment could reasonably be expected to have an adverse effect on product quality. The manufacturer shall ensure that maintenance and other personnel who are required to work temporarily under special environmental conditions are appropriately trained or supervised by a trained individual.

820.70(e) Contamination Control

Each manufacturer shall establish and maintain procedures to prevent contamination of equipment or product by substances that could reasonably be expected to have an adverse effect on product quality.

ANALYSIS

1   Personnel activities such as eating, drinking or smoking, which could have an adverse effect on a computer system, must be limited to a designated area away from the computer system. This must be specified in a written procedure.
2   There must be written procedures designed to prevent the contamination of the computer system.

820.70(g) Equipment

Each manufacturer shall ensure that all equipment used in the manufacturing process meets specified requirements and is appropriately designed, constructed, placed, and installed to facilitate maintenance, adjustment, cleaning, and use.

(1) *Maintenance schedule*. Each manufacturer shall establish and maintain schedules for the adjustment, cleaning, and other maintenance of equipment to ensure that manufacturing specifications are met. Maintenance activities, including the date and individual(s) performing the maintenance activities, shall be documented.

(2) *Inspection*. Each manufacturer shall conduct periodic inspections in accordance with established procedures to ensure adherence to applicable equipment maintenance schedules. The inspections, including the date and individuals conducting the inspections, shall be documented.

(3) *Adjustment*. Each manufacturer shall ensure that any inherent limitations or allowable tolerances are visibly posted on or near equipment requiring periodic adjustments or are readily available to personnel performing these adjustments.

ANALYSIS

1   Computer hardware/software must be properly specified to meet the requirements for the use for which it is intended.
2   Environmental controls, electrical requirements, control of electromagnetic 'noise,' etc., should be considered.
3   The location of the hardware must allow access for cleaning and maintenance.
4   There must be a written schedule for the maintenance of the computer hardware. This schedule should be readily available. A written record of all maintenance performed must be maintained.

820.70(i) Automated processes

When computers or automated data processing systems are used as part of production or the quality system, the manufacturer shall validate computer software for its intended use according to an established protocol. All software changes shall be validated before approval and issuance. These validation activities and results shall be documented.

ANALYSIS

1 There must be written procedures for the validation of computer systems.
2 Validation of computer systems must be performed by qualified individuals and must be verified.
3 There must be documented checks of input and outputs for accuracy.
4 Validation records must be maintained.
5 The failure of computer hardware and/or software within a finished device to meet its specifications must be investigated. The investigation must be documented and must include conclusions, corrective actions, and follow-up reviews.
6 There must be documented checks for programming errors.

820.72(b) Calibration

Calibration procedures shall include specific directions and limits for accuracy and precision. When accuracy and precision limits are not met, there shall be provisions for remedial action to reestablish the limits and to evaluate whether there was any adverse effect on the device's quality. These activities shall be documented.

(1) *Calibration standards.* Calibration standards used for inspection, measuring and test equipment shall be traceable to national or international standards. If national or international standards are not practical or available, the manufacturer shall use an independent reproducible standard. If no applicable standard exists, the manufacturer shall establish and maintain an in-house standard.
(2) *Calibration records.* The equipment identification, calibration dates, the individual performing each calibration and the next calibration date shall be documented. These records shall be displayed on or near each piece of equipment or shall be readily available to the personnel using such equipment and to the individuals responsible for calibrating the equipment.

ANALYSIS

1 All inspection, measuring or test equipment, whether mechanical, electrical, automatic, or mechanical in operation, which performs a quality assurance function must be validated for its intended use. This equipment must also be routinely calibrated according to a written procedure and the records that document these activities must be maintained.
2 For all software used in inspection, adequate and documented testing must validate the measuring and test equipment.
3 All hardware and/or software changes must be made by a designated individual(s) using a formal change control procedure.

4 Calibration procedures must include specific instructions and limits for accuracy and precision. They must also include procedures for remedial action if the limits are not met.

5 Personnel performing the calibrations must have the necessary education, training, experience, or background to perform the task. CVs should be kept on file.

6 The name of the calibrator, calibration date, and the date of the next calibration must be recorded. This information must be maintained by a designated individual(s) and must be readily retrievable for each piece of hardware requiring calibration.

SUB-PART H – ACCEPTANCE ACTIVITIES

820.80(b) Receiving Inspection and Testing

Each manufacturer shall establish and maintain procedures for acceptance of incoming product. Incoming product shall be inspected, tested, or otherwise verified as conforming to specified requirements. Acceptance or rejection shall be documented.

ANALYSIS

1 There must be a written procedure for the acceptance/rejection of computer hardware/software that is included in the finished device, and a designated individual(s) must carry out this process.

820.80 (d) Final Inspection and Testing

Each manufacturer shall establish and maintain procedures for finished device acceptance to ensure that each production run, lot, or batch of finished devices meets acceptance criteria. Finished devices shall be held in quarantine or otherwise adequately controlled until released. Finished devices shall not be released for distribution until:

(1) the activities required in the DMR are completed
(2) the associated data and documentation is reviewed
(3) the release is authorized by the signature of a designated individual(s), and
(4) the authorization is dated.

ANALYSIS

1 There must be written procedures for the evaluation of the finished device to assure that all computer hardware/software specifications have been met.

SUB-PART M – RECORDS

820.180 General Requirements

(b) *Records retention period* – All records required by this part shall be retained for a period of time equivalent to the design and expected life of the device, but in no case less than 2 years from the date of release for commercial distribution by the manufacturer.

ANALYSIS

1 Support documentation such as maintenance records, change control records, and project documentation should be retained as required by the Medical Devices GMPS (refer to 820.180(b)). This requirement is applicable to paper-based and electronic-based records.

2 The retention of supporting documentation such as maintenance records, change control records, and project documentation cannot be less than 2 years from the date of release of the medical device for commercial distribution by the manufacturer.

3 The retention of supporting documentation such as maintenance records, change control records, and project documentation depends on the following situations:

(a) obsolete system documentation — retained for the expected life of the device supported by the legacy functionality or system

(b) retired system documentation — retained for the expected life of the device supported by the retired system

(c) current documentation — retained permanently until (a) or (b) above apply

820.181 Device Master Record

Each manufacturer shall maintain device master records (DMRs). Each manufacturer shall ensure that each DMR is prepared and approved in accordance with Sec. 820.40. The DMR for each type of device shall include, or refer to the location of, the following information:

(a) Device specifications including appropriate drawings, composition, formulation, component specifications, and software specifications

(b) Production process specifications including the appropriate equipment specifications, production methods, production procedures, and production environment specifications

(c) Quality assurance procedures and specifications including acceptance criteria and the quality assurance equipment to be used

(d) Packaging and labelling specifications, including methods and processes used

(e) Installation, maintenance, and servicing procedures and methods

ANALYSIS

1 There should be an individual(s) with the responsibility for preparing, dating, and signing the master record.

2 The authorization of changes to the master records must be documented.

820.198 Complaint Files

(a) Each manufacturer shall maintain complaint flies. Each manufacturer shall establish and maintain procedures for receiving, reviewing, and evaluating complaints by a formally designated unit.

(b) Each manufacturer shall review and evaluate all complaints to determine whether an investigation is necessary. When no investigation is made, the manufacturer shall maintain a record that includes the reason no investigation was made and the name of the individual responsible for the decision not to investigate.

(c) Any complaint involving the possible failure of a device, labelling, or packaging to meet any of its specifications shall be reviewed, evaluated, and investigated, unless such investigation has already been performed for a similar complaint and another investigation is not necessary.

(d) Any complaint that represents an event that must be reported to FDA under part 803 or 804 of this chapter shall be promptly reviewed, evaluated, and investigated by a designated individual(s) and shall be maintained in a separate portion of the complaint flies or otherwise clearly identified. In addition to the information required by Sec. 820.198.

(e) When the manufacturer's formally designated complaint unit is located at a site separate from the manufacturing establishment, the investigated complaint(s) and the record(s) of investigation shall be reasonably accessible to the manufacturing establishment.

(f) If a manufacturer's formally designated complaint unit is located outside of the United States, records required by this section shall be reasonably accessible in the United States ... .

ANALYSIS

1 There should be a designated unit to review written and oral complaints.
2 The reasons for not investigating a complaint must be documented. The authorization of this decision must also be documented.
3 Complaints relating to injury, death or hazard to safety must be investigated immediately.
4 Records of an investigation must be maintained.

FOOD GMP REGULATIONS RELATED TO COMPUTER SYSTEMS VALIDATION

In the US, many regulations are applicable to food processing business. For example:

- Part 210, Good Manufacturing Practice Regulations
- Part 11, Electronic Records and Signatures
- Part 106, Infant Formula Quality Control Procedures
- Part 110, Current Good Manufacturing Practice in Manufacturing, Packing, and Holding Human Food
- Part 123, Fish and Fishery Products
- Part 129, Bottled Drinking Water

The US FDA is closely evaluating other areas such as food and dietary supplements, where proposed rulings have already been published (website address below).

A comprehensive US FDA guideline covering regulatory requirements applicable to computer systems performing regulated function in the food industry can be found at http://www.fda.gov/oralinspect_ref/igs/ foodcomp.html.

Appendix I

Technical Design Key Practices

For all software development, both large and small, the following design and programming techniques are consistent with current practices and may be used to facilitate the analysis of software components in order to reduce the chances of programming errors.

MODULAR DESIGN

- A modular design is recommended, especially for moderate- to large-scale software development efforts. Each software module should have well-defined and readily understood logical interfaces.
- Software components should be constructed using the principles of data abstraction (requirements decomposition). If available, a high-level, object-oriented language that supports the construction of abstract data types should be used.
- The software should be hierarchically structured as a series of layers.

SOFTWARE MODULE/PROCEDURE INTERFACES

- Entries to a software module or procedure should be through external calls on explicitly defined interfaces.
- Each procedure should have only one entry point and two exit points at most (one for normal exits and one for error exits).
- Data should be communicated between software modules and between procedures through the use of procedures except where necessary for the implementation of abstract data types. Input values should be checked for range errors using assertion statements (if provided by the programming language used).

INTERNAL CONSTRUCTION

- Each procedure should perform only a single, well-defined function.
- Control flow within a single thread of execution should be defined using only sequencing construct, with structured programming constructs being used for conditional routines (e.g., if-then-else or case routines), and structured constructs being used for loop routines (e.g., while-do or repeat-until routines).
- If concurrent execution is employed (e.g., via multiple threads, tasks, or processes), the software components should enforce limits on the maximum allowable degree of concurrency and should use structured synchronization constructs to control access to shared data.
- The equivalence of variables should not be used to permit multiple memory usage for conflicting purposes.
- Robust command parsing and range-checking mechanisms should be implemented to guard against malformed requests, out-of-range parameters, and I/O buffer overflows.

IN-LINE DOCUMENTATION

- Each software module, procedure, and major programming construct should be documented. The documentation should specify the functions performed along with a (formal or informal) specification of preconditions and post-conditions.
- Each loop should be preceded by a convincing argument (as a comment) that termination is guaranteed.
- Variable names should be used in only one context within the same procedure.
- Each variable should have an associated comment identifying the purpose of the variable and noting the range of allowable values, including if the range is unrestricted.
- If concurrency is employed, the documentation should specify how limits are enforced on the maximum allowable degree of concurrency and how accesses to shared data are synchronized in order to avoid (possibly undetected) run-time errors.

Index

A
Administrative procedures mapped to Part 11, 209–212
Administration procedures necessary to guard data integrity, confidentiality, and availability, 209–210
  physical safeguards, 211
  technical security services, 212
  technical security mechanisms necessary to guard against unauthorized access to data transmitted over a communication network, 213
Acquisition process, 147
  development process, 149–150
  maintenance process, 150–151
  operation process, 150
  supply process, 148
Applicability of a computer validation model, 167–172
  sample situation, 168–172

C
Change management, 87–89
  process, 88–89
Compliance Policy Guide (CPG) 7153.17, 125–127
  adequacy/timelines of the corrective action plan, 127
  impact on product quality/data integrity, 126
  nature/extent of the device, 125–126
Computer systems regulatory requirements, 219–238
  analysis, 219–238
Computer validation management cycle, 25–28
  validation guidelines, 26–28
  validation policies, 26
Computer validation program organization, 9–34

computer system validation executive committee, 30
computer system validation groups and teams, 31–32
CSV cross-functional team, 30–31
management group, 32
organizational model, 29
validation program coordinators, 32–34
Computer systems validation process, 35–41
  development life cycle and testing framework, 36
  system development files, 40–41
Criticality and complexity assessment, 173–181
  deliverables matrices, 177–181
  validation model decision tree, 174–177

E
Electronic record, 129–135
  audit trails applicable for electronic records, 131
  electronic record authenticity, 134–135
  events, 132–133
  how should Part 11 records be managed, 130–131
  instructions, 132
  minimum record retention requirements, 131
  preservation strategies, 133–134
  storage, 135
Electronic signatures, 137–140
  digital signatures, 138–140
  general concept, 137–138
  password-based signatures, 138

F
FDA, 13–19
  See also USA regulatory requirements for computer systems
Future, 153–155
  computer validation, 153–155

242

21 CFR Part 11: A Complete Guide to International Compliance

G
Glossary of terms, 157–164

H
hardware/software suppliers qualification, 107–109

I
Introduction 21 CFR Part 11, complete guide to international compliance, 1–3
Inspection and testing, 49–55
    black box testing, 52–54
    documents inspections and technical reviews, 50–51
    other testing types, 54–55
    regulatory guidance, 49–50
    white box testing, 51–52

M
Maintaining state of validation, 111–115
    business continuity, 114–115
    periodic review, 114
    ongoing verification program, 115
    operational management, 112–114
    security, 111–112

N
New computer systems validation model, 21–24

O
Operational checks, 121–124
Operation sequencing, 121–122
    Part 11 related operational checks, 122–124
    validation operational checks, 124

P
Part 11 remediation project, 117–120
    corrective action plan, 119
    evaluation of systems, 118
    remediation, 119–120

Q
Qualification, 57–80
    configurable software qualification, 76–78
    custom-built system qualification, 78–80
    hardware installation qualification , 58–61

operating system and utility software installation verification, 69–70
Part 11 areas of interest, 70, 72–73,75–76
related product for ISO/IEC 12119 IEEE standard adoption of ISO/IEC 12119, 73–75
software installation qualification, 61–64
standard instruments, micro controllers, smart instrumentation verification, 70–72
system operational qualification, 64–67
system performance qualification, 67–68
standard software packages qualification, 73

R
Relevant procedure controls, 85–86

S
Sample audit checklist for closed system, 215–218
    electronic signature and electronic records, 215–218
Sample development activities grouped by project periods, 183–208
    computer system incidents, 206–207
    conceptualization period, 183
    conduct a site acceptance test, 199–204
    development period, 183–195
    preparation of the qualification protocols, 195–199
    retirement, 204–206
    system life cycle, 208
Security, 93–104
    application security, 98–99
    network security, 97–98
    other key elements, 99–104
    physical security, 96
    security for computer system, 94–96
SLC documentation, 81–83
Source code, 105–106

T
Training, 91–92
    in the regulated industry, 91–92
Technical design key practices, 239–240
Technologies supporting Part 11, 141–146
    data encryption, 142–145
    digital signatures, 145–146

hash algorithms, 142
paper based versus electronic based
    solution, 141

U
USA regulatory requirements for computer
    systems, 13–17
    See FDA
    food industry, 18–19
    medical devices software, 17–18

V
Validation overview, 5–11
    computer systems validation for low-
        criticality or low complexity projects, 11

introduction to computer systems
    validation process, 9–11
key project elements, 8
what is a computer system, 5
what is a computer systems validation,
    5–6
which system should be validated, 8–9
why do we validate computer systems,
    6–7
Validation project plans and schedules,
    43–47
    mandatory signatures, 45
    project schedule, 45–47
    validation project plans, 43–45